MANAGING
CONSTRUCTION
EQUIPMENT

MANAGING CONSTRUCTION EQUIPMENT

S. W. NUNNALLY

North Carolina State University
Raleigh, North Carolina

PRENTICE-HALL, INC., Englewood Cliffs, New Jersey

Library of Congress Cataloging in Publication Data

NUNNALLY, S. W. (date)
    Managing construction equipment.

    Includes bibliographies and index.
    1. Construction equipment—Handbooks, manuals, etc.
I. Title.
TH900.N85   658.5   76–13527
ISBN 0–13–548339–5

Printed in the United States of America

10 9 8 7 6 5 4 3 2

PRENTICE-HALL INTERNATIONAL, INC., *London*
PRENTICE-HALL OF AUSTRALIA PTY. LIMITED, *Sydney*
PRENTICE-HALL OF CANADA, LTD., *Toronto*
PRENTICE-HALL OF INDIA PRIVATE LIMITED, *New Delhi*
PRENTICE-HALL OF JAPAN, INC., *Tokyo*
PRENTICE-HALL OF SOUTHEAST ASIA PTE. LTD., *Singapore*

To Joan, Steve, Jan, and John

CONTENTS

11

BITUMINOUS EQUIPMENT   **186**

12

EQUIPMENT ECONOMICS   **213**

13

MAINTENANCE, TIRES AND SAFETY   **233**

14

SYSTEM DESIGN AND SIMULATION   **244**

15
PLANT LAYOUT   261

INDEX   275

PREFACE

The objective of this book is to guide construction managers in planning, estimating, and directing construction equipment operations in a manner which will attain the best possible result. It is based on the author's more than twenty-five years experience in construction, engineering, and education. A special effort has been made to present such advanced management techniques as queueing theory and system simulation in a form which can be understood and used by those without a background in higher mathematics or operations research. Thus it is hoped that the material presented will be helpful to students in the areas of construction management and construction engineering as well as to those individuals already active in the construction industry.

A book of this type would not be possible without the assistance of many individuals and organizations. The cooperation of construction equipment manufacturers, the Construction Industry Manufacturers Association, and The Asphalt Institute in providing information and photographs and in permitting reproduction of certain of their material is gratefully acknowledged. Many of my former students and associates at the University of Florida have provided helpful comments and suggestions. I would particularly like to thank Professors James Schaub and Byron Ruth for their suggestions and for reviewing portions of the manuscript. Richard H. Small assisted greatly in the preparation of the computer programs presented in Chapter 14. Special thanks go to my wife, Joan, for her encouragement and for typing the manuscript.

Comments from readers regarding errors or possible improvements are earnestly solicited.

S. W. NUNNALLY

MANAGING CONSTRUCTION EQUIPMENT

1

INTRODUCTION— MATERIALS

1-1 INTRODUCTION

Since earliest time man the builder has sought to develop mechanical devices to facilitate his work in adapting the earth to serve the needs and desires of mankind. From the crude construction equipment utilized by ancient peoples has evolved the modern construction equipment used in building today's highways, airports, utility systems, factories, stores, and housing. Much of this construction equipment is utilized in the process of earthmoving—literally moving and reshaping the materials of the earth.

Gigantic as some of the construction equipment of today may seem, it will undoubtedly be dwarfed by the construction equipment of the future as man continues to seek faster and less expensive methods of construction. Those of you in the construction industry have the opportunity of contributing to the development of these new machines and methods which will be a part of tomorrow's construction world.

In spite of the rapid rise in labor and material costs in recent years, the cost of earthmoving has remained relatively stable because of the increasing productivity of the construction equipment involved. However, this increase in size and productivity of equipment also means that inefficiency and lost time in construction operations carry a correspondingly increased cost penalty. In a highly competitive

1

industry like the construction industry only those who achieve high levels of efficiency in planning and executing projects will survive and prosper.

The use of the planning, estimating, and operating techniques described in the following chapters should enable the construction equipment manager to attain high standards of performance. However, the construction equipment manager should also understand and apply a number of other planning and management techniques such as network planning methods (CPM/PERT), production and cost control methods, general principles of management, etc., to achieve the best possible results. Since there are a number of excellent references available on these subjects, they will not be covered in detail here.

Since much of the equipment to be discussed in this book is involved in the earthmoving process, it is important to have an understanding of the basic materials of earthmoving and their characteristics. The remainder of this chapter will be devoted to these topics.

1-2 MATERIALS OF EARTHMOVING

All of the materials found in the crust of the earth are properly classified as rock or soil. These then are the materials of interest to the earthmover. There are a number of properties of rock and soil that affect their behavior in the earthmoving process. Rock and its excavation and hauling will be discussed in Chapter 9. Soil properties are discussed below.

There are a number of soil characteristics that affect a soil's behavior for construction purposes. The designer and engineer are concerned with the ability of the soil to support a structure under the expected load and environmental conditions. The construction equipment manager is concerned with such items as the soil's trafficability, loadability, and moisture content as well as its drainage, weight, and volume change characteristics.

Laboratory tests are required to fully determine a soil's physical and chemical properties which control those characteristics mentioned above. However, the ability to identify major soil types and a knowledge of their behavior characteristics will greatly aid the equipment manager in the proper choice of equipment and the selection of the method of equipment operation for a particular job. Basic soil types and field identification methods are presented in Section 1-5.

Trafficability

The ability of a soil to support the load of wheeled or tracked vehicles under repeated traffic is called *trafficability*. Trafficability during construction operations is clearly important to the construction equipment manager. Trafficability is usually estimated based on the soil type and moisture conditions expected during construction or it is determined by test operations. However, devices for direct measurement of a soil's trafficability are available. Poor trafficability conditions may require special measures for drainage, the use of low ground pressure construction equipment, stabilization of haul routes, or a combination of these measures for efficient equipment operations.

Loadability

Loadability is a subjective measure of the ease with which a soil can be excavated and loaded. Generally loose, granular, noncohesive soils are very loadable. Compacted cohesive soils and most rocks have a low loadability.

Moisture Content and Drainage Characteristics

All soils in the natural state contain some moisture. The effect of moisture on the compaction process will be discussed in Chapter 7. The moisture content of soil also affects the soil's weight and handling characteristics. The moisture content of a soil is expressed as the percentage of the soil's dry weight which equals the weight of water in the soil. For example, a soil sample weighing 165 lb in the natural state weighs 150 lb after drying. The weight of water in the sample was therefore 15 lb. The moisture content is then found to be

$$\frac{15}{150} \times 100 = 10\%$$

The soil is thus described as having a 10% moisture content. Soil drainage characteristics are important to trafficability in addition to influencing the moisture content and weight of materials to be excavated and hauled. Drainage characteristics by soil type are indicated in Table 1-5.

Weight

The weight per unit volume of a soil is determined by the soil type, its degree of compaction, and its moisture content. The dry density (dry weight per unit volume, usually expressed in pounds per cubic foot) of a soil is frequently used as a measure of the soil's degree of compaction. This will be discussed further in Chapter 7. There is also a relationship between the dry density of a soil and its strength and bearing capacity. Soil weight may be a limiting factor in determining the capacity of an excavator or haul unit. Some typical weights of common soils are given in Table 1-1.

TABLE 1-1 *Approximate material characteristics**

Material	Loose (lb/cu yd)	Bank (lb/cu yd)	Swell (%)	Load Factor
Clay, dry	2,100	2,650	26	0.79
Clay, wet	2,700	3,575	32	0.76
Clay and gravel, dry	2,400	2,800	17	0.85
Clay and gravel, wet	2,600	3,100	17	0.85
Earth, dry	2,215	2,850	29	0.78
Earth, moist	2,410	3,080	28	0.78
Earth, wet	2,750	3,380	23	0.81
Gravel, dry	2,780	3,140	13	0.88
Gravel, wet	3,090	3,620	17	0.85
Sand, dry	2,600	2,920	12	0.89
Sand, wet	3,100	3,520	13	0.88
Sand and gravel, dry	2,900	3,250	12	0.89
Sand and gravel, wet	3,400	3,750	10	0.91

*Exact values will vary with grain size, moisture content, compaction, etc. Test to determine exact values for specific soils.

1-3 VOLUME CHANGE CHARACTERISTICS OF SOILS

Soil States

Following are the three principal states in which earthmoving material may exist:

Natural (in-place):   Soil in its natural state. A unit volume of material is referred to as a *bank cubic yard* (BCY) or *a bank cubic meter* (Bm³).

Loose: Soil after excavation or loading. A unit volume is referred to as a *loose cubic yard* (LCY) or *a loose cubic meter* (Lm³).

Compacted: Soil after compaction. A unit volume is referred to as a *compacted cubic yard* (CCY) or *a compacted cubic meter* (Cm³).

When a soil is excavated, it typically increases in volume so that the weight of soil contained in one bank cubic yard will, after excavation, occupy more than one cubic yard. This increase in volume due to excavation is called *swell*. An equation for determining swell is

$$\text{swell } (\%) = \left(\frac{\text{weight/bank cubic yard}}{\text{weight/loose cubic yard}} - 1 \right) \times 100 \qquad (1\text{-}1)$$

Similarly, when a soil is compacted, it reduces in volume so that a given weight of soil will occupy somewhat less volume than it did in its natural state. This process is called *shrinkage*. An equation for determining shrinkage is

$$\text{shrinkage } (\%) = \left(1 - \frac{\text{weight/bank cubic yard}}{\text{weight/compacted cubic yard}}\right) \times 100 \qquad (1\text{-}2)$$

It is important in earthmoving calculations to convert all volumes to a common unit of measure. The bank cubic yard is commonly used for this purpose, although any of the three volume units might be chosen. The volume unit specified as the basis for payment in an earthmoving contract is often referred to as a *pay yard*. A pay yard may be any of the three units referred to above.

To simplify conversion of loose cubic yards to bank cubic yards, the term *load factor* is often used. The load factor may be calculated as follows:

$$\text{load factor} = \frac{\text{weight/loose cubic yard}}{\text{weight/bank cubic yard}} \qquad (1\text{-}3)$$

or

$$\text{load factor} = \frac{1}{1 + \text{swell}} \qquad (1\text{-}4)$$

Hence, loose volume multiplied by the load factor yields bank volume. Typical values of swell and load factor for some common soils are given in Table 1-1. Approximate soil volume conversion factors are given in Table 1-2. Soil volume change due to excavation and compaction is illustrated in Figure 1-1.

TABLE 1-2 *Typical soil volume conversion factors*

Soil Type	Initial Soil Condition	Converted to:		
		Bank	Loose	Compacted
Clay	Bank	1.00	1.27	0.90
	Loose	.79	1.00	.71
	Compacted	1.11	1.41	1.00
Common earth	Bank	1.00	1.25	0.90
	Loose	.80	1.00	0.72
	Compacted	1.11	1.39	1.00
Rock (blasted)	Bank	1.00	1.50	1.30
	Loose	0.67	1.00	0.87
	Compacted	0.77	1.15	1.00
Sand	Bank	1.00	1.12	0.95
	Loose	0.89	1.00	0.85
	Compacted	1.05	1.18	1.00

Example 1-1

PROBLEM: What are the conversion factors for loose sand which is initially in (a) the bank state, (b) the loose state, or (c) the compacted state?

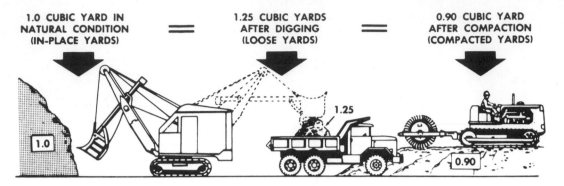

| 1.0 CUBIC YARD IN NATURAL CONDITION (IN-PLACE YARDS) | = | 1.25 CUBIC YARDS AFTER DIGGING (LOOSE YARDS) | = | 0.90 CUBIC YARD AFTER COMPACTION (COMPACTED YARDS) |

Figure 1-1 Typical volume change due to handling. (U.S. Department of the Army[8].)

Solution (Table 1-2):

1.0 BCY = 1.12 LCY = 0.95 CCY.

1.0 LCY = 0.89 BCY = 0.85 CCY.

1.0 CCY = 1.05 BCY = 1.18 LCY.

Example 1-2

PROBLEM: If 1 LCY of cinders weighs 1,000 lb and 1 BCY of the same material weighs 1,600 lb, what is the load factor for this material?

Solution:

$$\text{Load Factor} = \frac{\text{weight/LCY}}{\text{weight/BCY}} = \frac{1,000}{1,600} = 0.625.$$

1-4 SPOIL BANK DIMENSIONS

Basic Considerations

In earthmoving situations it will sometimes be necessary to calculate the size of spoil banks which will be created by the material from an excavation. Since the material in the spoil bank will, of course, be in the loose state, it is necessary to first convert the volume of excavation from in-place conditions (BCY) to loose conditions (LCY). The size of spoil bank which will contain this volume can then be obtained from tables such as Table 1-3 or may be calculated as described below.

The *angle of repose* is the natural angle that the sides of a spoil bank will form with the horizontal when the excavated soil is dumped onto the spoil bank. This represents the equilibrium position of the soil and will vary with the physical characteristics of the soil. Typical angles of repose for common soils are given in Table 1-3.

TABLE 1-3 *Spoil bank dimensions**

Material	Angle of Repose
Dry sand	$30° = 1\frac{3}{4}$ to 1
	$33.6° = 1\frac{1}{2}$ to 1
Gravel	$35° = 1\frac{7}{16}$ to 1
Clay	$35° = 1\frac{7}{16}$ to 1
Earth	$37° = 1\frac{5}{16}$ to 1
Moist sand	$37° = 1\frac{5}{16}$ to 1

Area spoil bank = Area excavation + 20%

Width of Base Feet	$30° = 1\frac{3}{4}$ to 1			$33.6° = 1\frac{1}{2}$ to 1			$37° = 1\frac{5}{16}$ to 1			$45° = 1$ to 1		
	Height Ft.–In.	Spoil Area Sq. Ft.	Equiv. Excav. Area Sq. Ft.	Height Ft.–In.	Spoil Area Sq. Ft.	Equiv. Excav. Area Sq. Ft.	Height Ft.–In.	Spoil Area Sq. Ft.	Equiv. Excav. Area Sq. Ft.	Height Ft.–In.	Spoil Area Sq. Ft.	Equiv. Excav. Area. Sq. Ft.
16	4– 8	37	30.8	5–4	43	35.8	6– 1	48	40.0	8–0	64	53.3
17	4–10	42	35.0	5–8	48	40.0	6– 5	54	45.0	8–6	72	60.0
18	5– 2	47	39.1	6–0	54	45.0	6– 9	61	50.8	9–0	81	67.5
19	5– 6	52	43.3	6–4	60	50.0	7– 2	68	56.6	9–6	90	75.0
20	5– 9	58	48.0	6–8	66	55.0	7– 6	75	62.4	10–0	100	83.3
21	6– 1	64	53.3	7–0	74	61.5	7–10	83	69.0	10–6	110	92.0
22	6– 4	70	58.3	7–4	81	67.5	8– 3	91	75.8	11–0	121	101
23	6– 7	76	63.3	7–8	88	73.0	8– 8	99	87.5	11–6	132	110
24	6–11	83	69.0	8–0	96	80.0	9– 1	109	91.0	12–0	144	120
25	7– 2	90	75.0	8–4	104	87.0	9– 5	117	97.2	12–6	156	130
26	7– 6	98	81.3	8–8	112	93.5	9–10	126	105	13–0	169	141
27	7– 9	104	86.5	9–0	122	101	10– 0	135	112	13–6	182	152
28	8– 1	113	94.2	9–4	130	108	10– 7	148	123	14–0	196	163
29	8– 5	122	102	9–8	140	117	11– 0	158	131	14–6	210	175
30	8– 7	129	107	10–0	150	125	11– 4	170	142	15–0	225	187
32	9– 3	148	123	10–8	170	141	12– 1	193	160	16–0	256	213
34	9–10	167	139	11–4	193	161	12–10	218	182	17–0	289	240
36	10– 5	187	156	12–0	216	180	13– 1	244	208	18–0	324	270
38	11– 0	208	173	12–8	240	200	14– 4	272	227	19–0	361	300
40	11– 7	231	192	13–4	266	221	15– 1	301	250	20–9	400	333
42	12– 1	255	212	14–9	294	244	15–10	332	276	21–0	441	367
44	12– 8	279	232	14–8	323	269	16– 1	365	304	22–0	484	402
46	13– 3	305	254	15–4	335	279	17– 4	399	332	23–0	529	440
48	13–10	333	278	16–0	384	320	18– 1	434	361	24–0	576	480
50	14– 5	361	300	16–8	416	348	18–10	471	393	25–0	625	520
52	15– 0	390	324	17–4	450	375	19– 7	509	424	26–0	676	563
54	15– 7	421	350	18–0	486	405	20– 4	549	457	27–0	729	605
56	16– 2	453	378	18–8	523	436	21– 1	591	492	28–0	784	650
58	16– 9	486	404	19–4	562	470	21–10	634	528	29–0	841	700
60	17– 4	520	433	20–0	600	500	22– 7	678	565	30–0	900	750
62	17–11	555	463	20–8	642	565	23– 4	734	603	31–0	961	800

Calculating Spoil Bank Dimensions

Longitudinal spoil banks typically have a triangular cross section (as illustrated in Table 1-3). Spoil piles have a conical shape. This permits the bank volume to be easily calculated by trigonometric relations. Conversely, the size of bank required to hold a specified amount of spoil may be found. Equations may be developed as follows, where R equals the angle of repose of the soil.

For a Triangular Bank

$$\text{volume} = \text{area} \times \text{length}$$

$$\text{cross section area} = \frac{\text{base width } (B)^2}{4} \times \text{tangent } R \tag{1-5}$$

$$\text{height} = \frac{B}{2} \times \text{tangent } R \tag{1-6}$$

For a Circular Pile

$$\text{volume} = \tfrac{1}{3} \times \text{base area} \times \text{height}$$

$$\text{volume} = 0.131 \times \text{diameter of base } (D)^3 \times \tan R \tag{1-7}$$

$$\text{height} = \frac{D}{2} \times \tan R \tag{1-8}$$

Example 1-3

PROBLEM: Find the base width and height of a spoil bank that is created when excavating a ditch having a cross-section area of 100 sq ft. The soil is common earth having a swell of 25% and an angle of repose of 37°.

Solution:

For each linear foot excavated, the volume of excavation removed is 100 bank cu ft. Thus,

Loose volume per linear foot $= 100 \times (1 + S_w) = 100 \times 1.25 = 125$ cu ft.

Area of spoil bank per linear foot $= 125$ sq ft.

$125 = \tfrac{1}{4} \times (B)^2 \times \tan 37°$ (Equation 1-5).

$(B)^2 = 4 \times 125 \div 0.7536 = 663.5$ sq ft.

Base width $(B) = (663.5)^{1/2} = 25.8$ ft.

Height $= \dfrac{B}{2} \times \tan 37° = \dfrac{25.8}{2} \times 0.7536 = 9.7$ ft (Equation 1-6).

Example 1-4

PROBLEM: Find the size of a circular spoil pile that will contain 100 BCY of sand excavation. Use the volume conversion factors in Table 1-2. The angle of repose of the soil is 32°.

Solution:

Volume of pile $= 100 \times 1.12 = 112$ LCY $\times 27 = 3{,}024$ loose cu ft (Table 1-2).

$3{,}024 = 0.131 \times (D)^3 \times \tan 32°$ (Equation 1-7).

$(D)^3 = 3024 \div (0.131 \times 0.6249) = 36{,}940$ ft^3.

Base diameter $(D) = (36{,}940)^{1/3} = 33.3$ ft.

Height $= \dfrac{D}{2} \times \tan 32° = \dfrac{33.3}{2} \times 0.6249 = 10.4$ ft (Equation 1-8).

1-5 SOIL CLASSIFICATION AND IDENTIFICATION

Classification Systems

There are two principal soil classification systems used in the United States. These are the Unified Soil Classification System used by many engineers and government agencies and the AASHO (American Association of State Highway Officials) System widely used for highway design and construction. In both of these systems soil is considered to be composed of five fundamental soil groups: gravel, sand, silt, clay, and organic material. Although the size ranges for these basic soil groups vary slightly for the two systems, both systems classify a soil according to the amount and type of each basic soil included. Other principal factors used for soil classification include the shape of the grain-size distribution curve as well as the plasticity and compressibility characteristics of the soil. Classification by the Unified System will be described below. Soil behavior and suitability for various construction purposes may be predicted by using either system.

The Unified Soil Classification System

The Unified Soil Classification System uses a symbol that consists of two letters to describe a soil. After removing particles over 3 in. in diameter (classified as cobbles), soils are classified as coarse-grained (less than 50% passing the No. 200 sieve), fine-grained (50% or more passing the No. 200 sieve), or organic (containing a high percentage of fibrous organic matter). Further classification is described below.

Coarse-grained Soils

The first letter of the classification symbol identifies the predominant fraction of the material retained on the No. 200 sieve. Gravel (G) is material that passes the 3-in. sieve but is retained on the No. 4 sieve. Sand (S) is material that passes the No. 4 sieve but is retained on the No. 200 sieve. Thus, if 50% or more of the coarse material is retained on the No. 4 sieve, the soil is assigned the letter G. Otherwise, the letter S is assigned.

The second letter is determined by the percentage and type of material passing the No. 200 sieve. If less than 5% of the total sample passes the No. 200 sieve, the soil is classified as either well-graded (W) or poorly graded (P). Gradation is determined by the shape of the grain-size distribution curve. A poorly graded soil will have a small difference in size between the largest and smallest grains or will have a deficiency in one or more sizes of particles. Typical gradation curves for these soils are illustrated in Figure 1-2. These soils are thus classified as GW, GP, SW, or SP.

When the portion of the sample passing the No. 200 sieve exceeds 12%, the second letter is assigned according to the plasticity and compressibility (as measured by the liquid limit and plasticity index) of the fraction passing the No. 40 sieve. The liquid limit (LL) of a soil is the water content expressed in percentage of dry weight at which a soil will start to flow when tested in a standard laboratory test device. The liquid limit corresponds to the moisture content at which the soil passes from the plastic state into the liquid state. The plastic limit (PL) of a soil is the moisture content (in percent) at which it begins to crumble when rolled into a thread $\frac{1}{8}$ in. in diameter. The plastic limit corresponds to the moisture content at which the soil passes from the plastic state into the semi-solid state. The plasticity index (PI) is the numerical difference between the liquid limit and the plastic limit. Thus, the plasticity index corresponds to the range in moisture content over which the soil remains in the plastic state. A combination of liquid limit and plasticity index is used to determine a soil's plasticity. If the fine material exhibits little or no plasticity, it is considered silt-like (M). If the fines exhibit plasticity, they are classified as clay-like (C). These soils will thus be classified as GC, GM, SC, or SM. If the percentage of the sample passing the No. 200 sieve falls between 5% and 12%, a dual symbol that combines the two methods above is used. These soils would then be classified as GW–GM, etc.

Fine-grained Soils

Soils having more than 50% passing the No. 200 sieve are classified by plasticity and compressibility rather than by grain size. Liquid limit and plastic limit tests are run on the portion passing the No. 40 sieve to permit classification.

When the liquid limit is less than 50%, the soil is classified as having a low compressibility and the second letter will be L. The first letter will be assigned based on a plot of plasticity index and liquid limit values. Such a plot will identify the predominant material as clay (C) or silt (M). If organic material is identified by color, odor, or a radical drop in liquid and plastic limits after drying, the material is classified as organic (O). Classification symbols for these soils then may be CL, ML, or OL.

When the liquid limit is 50% or greater, the soil is classified as being highly compressible and the second letter assigned is H. The first letter is assigned in the

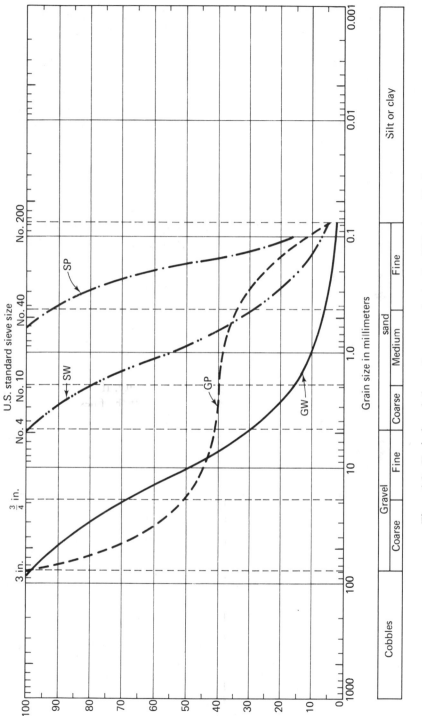

Figure 1-2 Typical gradations of coarse-grained soils. (U.S. Army Engineer School.)

same manner as for low compressibility soils discussed above. Classification symbols for these soils may thus be CH, MH, or OH.

Organic Soils

Soils that contain a large percentage of fibrous organic matter (such as peat and partially decomposed vegetation) are classified by the symbol Pt. Such soils are usually readily identified by their color (dull brown to black), odor, spongy feel, or fibrous texture.

Field Identification Procedures

When time and facilities do not permit laboratory testing of a soil, the use of the procedures described below in conjunction with Table 1-4 will permit a reasonably accurate soil classification to be made. All particles over 3 in. in diameter are

TABLE 1-4   *Field identification procedures (Unified System)*

Symbol	Name	Field Identification Procedure*		
Coarse-grained soils (less than 50% pass No. 200)		Portion of coarse fraction less than $\frac{1}{4}$ in. diameter: Gravel = less than 50% $\qquad$ Sand = 50% or more		
GW SW	Well-graded gravel Well-graded sand	Wide range of grain sizes with all intermediate sizes		
GP SP	Poorly graded gravel Poorly graded sand	Predominately one size or some intermediate sizes missing		
GM SM	Silty gravel Silty sand	Fines with low plasticity (see ML below)		
GC SC	Clayey gravel Clayey sand	Plastic fines (see CL below)		
Fine-grained soils (50% or more pass No. 200)		Tests on fraction smaller than No. 40 (approximately $\frac{1}{64}$ in.)		
		Dry Strength	*Shaking*	*Toughness*
ML	Low plasticity silt	Low	Quick to slow	None
CL	Low plasticity clay	Medium	None to slow	Medium
OL	Low plasticity organic	Low to medium	Slow	Slight
MH	High plasticity silt	Low to medium	None to slow	Slight to medium
CH	High plasticity clay	High	None	High
OH	High plasticity organic	Medium to high	None to slow	Slight to medium
PT	Peat or organic	Color, odor, spongy feel, fibrous texture		

*See text for discussion of test procedures.

first removed. If more than half the sample by weight is larger than the No. 200 sieve (the No. 200 sieve is approximately the smallest particle that can be seen by the naked eye), the soil is classified as coarse-grained. The sample is then separated at the No. 4 sieve size (approximately $\frac{1}{4}$ in). If 50% or more of the coarse fraction is larger than the No. 4 sieve, the soil is predominantly gravel; otherwise, it is sand. Further classification is then made in accordance with Table 1-4. For classifying fine-grained soils, only the portion passing the No. 40 sieve (approximately $\frac{1}{64}$ in.) is used. Procedures for making the three tests identified in Table 1-4 are as follows.

Dry Strength

Mold a sample approximately $1\frac{1}{2}$ in. in diameter to the consistency of putty; add water if necessary. Allow the sample to dry thoroughly. Break the sample by using the thumb and forefingers of both hands and try to powder it by rubbing it with the thumb and forefinger of one hand. Samples of very highly plastic soils cannot be broken or powdered in these tests. Since samples of nonplastic soils, however, will have little or no dry strength, they will crumble and powder upon being picked up.

Shaking

Form a ball of material approximately $\frac{3}{4}$ in. in diameter; add water until it is just wet enough not to stick to the fingers as it is molded. Place the sample in the palm of the hand and shake vigorously. Reaction to this test is judged by the speed with which water comes to the surface of the sample and produces a shiny appearance. A rapid reaction is typical of nonplastic silts. However, even a slight amount of colloidol clay in the sample will greatly slow up the reaction.

Toughness (Roll Test)

Form a ball of material in the same manner as described for the shaking test. Then roll the sample on a nonabsorbent surface until a thread approximately $\frac{1}{8}$ in. in diameter is formed. Fold the thread and repeat until the thread crumbles; this is the plastic limit. The tougher the thread near the plastic limit, the higher the colloidal clay content. A highly plastic soil may be remolded into a ball and the ball deformed without cracking or crumbling. A soil of low plasticity cannot be remolded into a ball without breaking up.

Characteristics of Soil Types

The general characteristics of the soil types discussed above which are significant to construction equipment operations are shown in Table 1-5.

TABLE 1-5 *Soil characteristics (Unified System)*

Symbol	Drainage Characteristics	Workability as Construction Material	Suitability for Subgrade		Suitability as Surfacing
			No Frost Action	Frost Action	
GW	Excellent	Excellent	Excellent	Excellent	Good
GP	Excellent	Good	Good to excellent	Good	Poor
GM	Fair to poor	Good	Good to excellent	Fair to good	Fair
GC	Poor	Good	Good	Fair	Excellent
SW	Excellent	Excellent	Good	Fair to good	Good
SP	Excellent	Fair	Fair to good	Fair	Poor
SM	Poor to fair	Fair	Fair to good	Poor to good	Fair
SC	Poor	Good	Poor to fair	Poor	Excellent
ML	Poor to fair	Fair	Poor to fair	Unsuitable	Poor
CL	Very poor	Fair to good	Poor to fair	Unsuitable	Fair
OL	Poor	Fair	Poor	Unsuitable	Poor
MH	Poor to fair	Poor	Poor	Unsuitable	Poor
CH	Very poor	Poor	Poor to fair	Unsuitable	Poor
OH	Very poor	Poor	Poor to very poor	Unsuitable	Poor
Pt	Poor to fair	Unsuitable	Unsuitable	Unsuitable	Unsuitable

PROBLEMS

1. A soil weighs 2,000 lb/cu yd loose and 2,500 lb/cu yd in-place. Find the swell and load factor for the soil.

2. A scraper has a load capacity of 100,000 lb and a heaped volume capacity of 35 cu yd. How many bank cubic yards of the soil in Problem 1 can the scraper carry?

3. A soil weighs 1,920 lb/cu yd loose, 2,400 lb/cu yd in-place and 2,825 lb/cu yd compacted. Find the swell and shrinkage of this soil.

4. A project requires that 100,000 in-place cu yd of the soil in Problem 3 be excavated, hauled, and compacted in a fill. How many loose cubic yards must be hauled? How many compacted cubic yards of fill will the excavation yield?

5. A ditch having a cross-sectional area of 100 sq ft is being excavated in clay. Using Equations 1-5 and 1-6, find the height and width of spoil bank which will result. The soil's angle of repose is 35° and its swell is 40%. Compare the results with Table 1-3 and explain the difference.

6. Find the size of a circular stockpile resulting from the excavation of 1,000 BCY of gravel. Swell is 12% and angle of repose is 35°.

7. Using Table 1-2 for volume conversion, find the size of a spoil bank 100 ft long resulting from the excavation of 1,000 cu yd of sand. The angle of repose is 30°.

8. Classify the following soil. Field identification test results are:
   (a) More than 50% by weight is larger than No. 200 sieve.
   (b) More than half the coarse fraction is larger than $\frac{1}{4}$ in.
   (c) There is an appreciable amount of fines present. Tests show these to be non-plastic.

9. Classify the soil whose field identification test results are shown below.
   (a) More than 50% by weight is smaller than No. 200 sieve.
   (b) Dry strength is high.
   (c) Shaking test gives no reaction.
   (d) Toughness is high.

10. You have identified a soil as silt of low plasticity (ML). What would you expect the drainage and workability characteristics of this soil to be?

REFERENCES

1. ADRIAN, JAMES J., *Quantitative Methods in Construction Management.* New York: American Elsevier, 1973.

2. *Basic Estimating* (3rd ed.). Melrose Park, Illinois: Construction Equipment Division, International Harvester Company, n.d.

3. *Cost Control and CPM in Construction.* Washington, D.C.: The Associated General Contractors of America, 1968.

4. *Fundamentals of Earthmoving.* Peoria, Illinois: Caterpillar Tractor Company, 1975.

5. O'BRIEN, JAMES J., *CPM in Construction Management.* New York: McGraw-Hill, 1971.

6. *Soils Manual for Design of Asphalt Pavement Structures.* College Park, Maryland: The Asphalt Institute, 1963.

7. TERZAGHI, KARL, and RALPH B. PECK, *Soil Mechanics in Engineering Practice* (2nd ed.). New York: John Wiley & Sons, 1967.

8. *TM 5-331A: Earthmoving, Compaction, Grading and Ditching Equipment.* Washington, D.C.: U.S. Department of the Army, 1967.

2

POWER CRANES
AND EXCAVATORS

2-1 THE CRANE-SHOVEL AND ITS ATTACHMENTS

Introduction

The crane-shovel with its variety of attachments was among the earliest pieces of modern construction equipment and is still the most common lifting and loading equipment used in construction. William S. Otis is usually credited with developing the first power shovel when in 1836 he created a machine which mechanically duplicated the motions of a man digging with a hand shovel.

Components of a Crane-shovel

A crane-shovel is made up of three basic parts: the carrier (or mounting), a revolving superstructure (also called revolving deck or turntable), and a front-end attachment. There are three types of carriers or mountings available as shown in Figure 2-1.

Carriers

A crawler mounting provides a stable platform for the revolving superstructure and affords excellent mobility on the job site while exerting a low ground pressure. Its low ground pressure (approximately 5 to 10 lb/sq in.) makes this the preferred

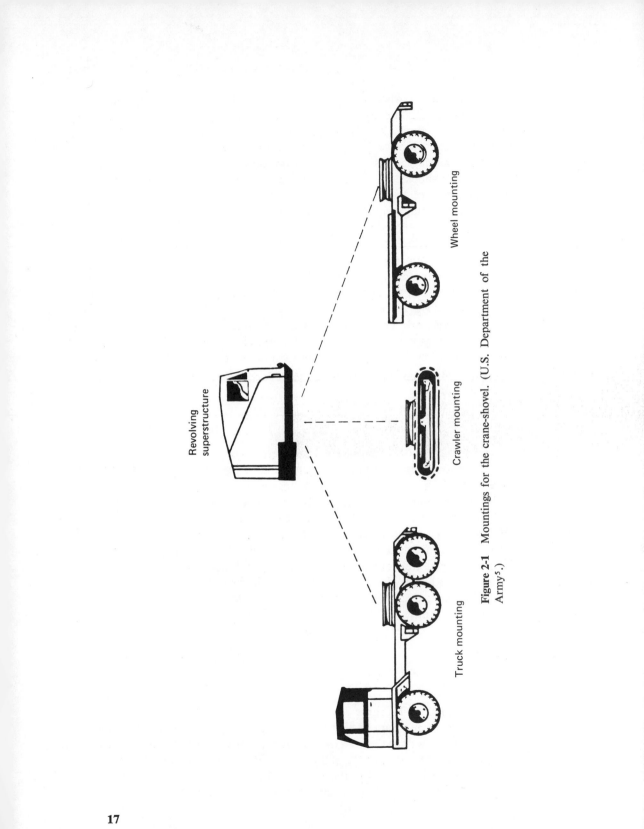

Figure 2-1 Mountings for the crane-shovel. (U.S. Department of the Army[5].)

Revolving superstructure

Wheel mounting

Crawler mounting

Truck mounting

type of mounting whenever poor soil trafficability conditions are encountered. Extra wide tracks are available whenever especially low ground pressures are required. The crawler treads may be flat when traction is not a problem or they may be tractor-type (or grouser) treads when increased traction is required. The crawler mounting is also used for heavy and rock excavation when increased resistance to shock and tread wear is needed. Because of their low travel speed (usually 2 mph or less), crawler mounted crane-shovels should be transported by special heavy equipment trailers when travel distance exceeds approximately one mile.

Rubber-tired (truck or wheel) mountings enable the crane-shovel to transport itself rapidly from job to job over highways, but they provide a less stable base than a crawler mounting. A rubber-tired mounting may also be preferred whenever very abrasive work surfaces would cause track damage or whenever crawler tracks are prohibited (for example, on paved roads).

The primary difference between truck mountings and wheel mountings is that truck mountings use separate engines for vehicle propulsion and crane operation and wheel mountings use one engine for both purposes. Truck mountings allow highway speeds up to approximately 50 mph. Wheel mountings are usually limited to 30 mph or less.

Attachments

Figure 2-2 shows the principal members of the crane-shovel family. The name of a particular piece of equipment belonging to the crane-shovel family is determined by the front-end attachment being used. Thus, a crane-shovel being used with a dragline attachment is usually referred to as simply a dragline. However, there has been a recent trend toward production of specialized equipment which performs only one of the functions shown in Figure 2-2. For example, a number of pieces of equipment are available to operate only as hoes (also called backhoes or excavators). Generally, the crane-shovel and its attachments may be either cable operated or hydraulically powered. The current trend is toward increased use of hydraulic equipment. Characteristics and application of each member of the crane-shovel family will be covered in the sections which follow. Other types of excavators and loaders will be presented in Chapter 3.

Bucket Volume

The rated bucket capacity for members of the crane-shovel family is determined as indicated in Table 2-1. Since most of these ratings are based on struck volume, it is often assumed that the heaping of material in the bucket will compensate for the swell of the material during excavation. Thus, a 3-cu yd shovel bucket would be considered capable of holding 3 BCY of material. For an improved estimate of bucket capacity, the rated bucket capacity should be multiplied by a *bucket fill factor* or *bucket efficiency factor*. Some suggested values of bucket fill

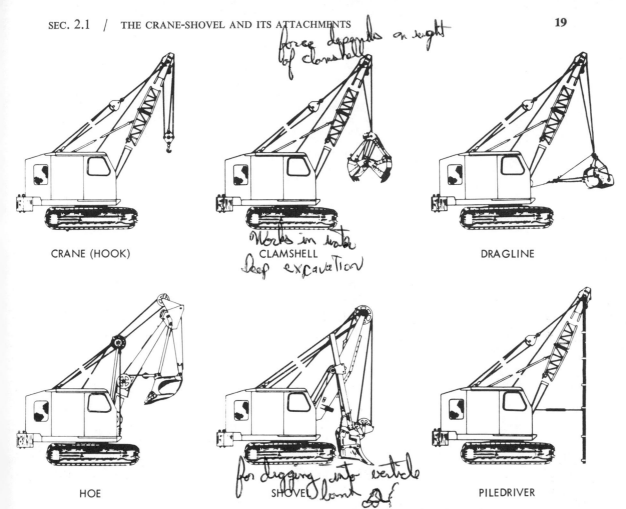

Figure 2-2 Crane-shovel attachments. (U.S. Department of the Army[5].)

TABLE 2-1 *Method of rating bucket capacity*

Attachment	Rated Bucket Capacity
Clamshell	Plate line capacity or water line capacity
Dragline	90% of struck volume
Hoe	Struck volume
Shovel	Struck volume

factors are given in Table 2-2. Thus, for common earth having a bucket fill factor of 0.85, a 3-cu yd shovel bucket would be expected to hold an average of $3 \times 0.85 = 2.55$ BCY per load. When the heaped bucket volume is known, a more accurate estimate of bank cubic yards per bucket load is found by multiplying the heaped volume by the load factor by the bucket efficiency factor.

TABLE 2-2 *Bucket fill factors for excavators*

Material	Bucket Fill Factor
Sand and gravel	.90–1.00
Common earth	.80– .90
Hard clay	.65– .75
Wet clay	.50– .60
Rock, well-blasted	.60– .75
Rock, poorly blasted	.40– .50

2-2 CRANES

General

The crane-shovel equipped with a crane boom and hook as illustrated in Figure 2-2 is referred to simply as a *crane*. Cranes are used primarily for lifting a load, moving it horizontally by swinging or traveling, and then lowering or dumping it into the desired position. In addition to the hook already illustrated, there are a number of special-purpose tools available for use with the crane as seen in Figure 2-3. The use of digging tools and the pile driver will be covered in later sections of this chapter.

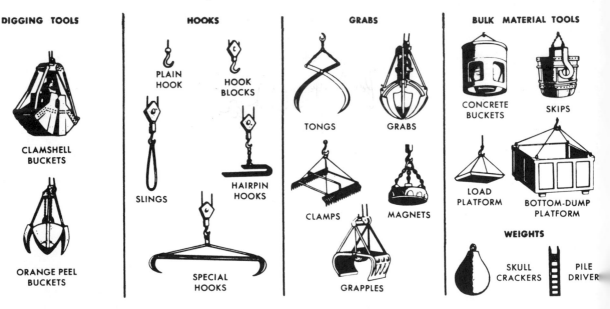

Figure 2-3 Crane boom attachments. (Permission to reproduce this material has been granted by the Construction Industry Manufacturers Association (CIMA). CIMA assumes no responsibility for the accuracy of this reproduction.)

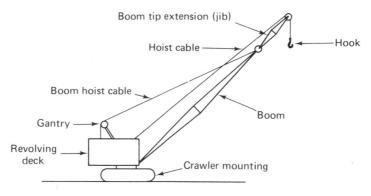

Figure 2-4   Nomenclature of crane components.

The nomenclature of crane components is illustrated in Figure 2-4. Even though the crane boom, revolving deck, and carrier may be hydraulically operated, a cable (wire rope) is normally used in conjunction with the hook for lifting and lowering operations. A large (130-ton) hydraulic truck crane is shown in Figure 2-5.

Figure 2-5   Large hydraulic truck crane. (Courtesy Harnischfeger Corp.)

Lifting Capacity

The safe lifting capacity of a crane is determined primarily by its operating radius, that is, the horizontal distance from the center of rotation of the super-structure to the hook. This distance is a function of the boom length and boom angle above the horizontal. Other factors which affect safe load capacity include: type of working surface, use of outriggers, amount of counterweight, size and type of wire rope used, size of hook block, position of the boom in relation to the carriage, and the maintenance condition of the crane.

Manufacturers provide capacity charts to assist in determining the safe load capacity of their cranes under specific conditions. Electronic load indicators are also available that show the actual load on the boom at any time and that warn when the safe load capacity is being approached. However, it is well to understand the method for classifying crane capacity and the use of the term *tipping load*.

A standard method for rating cranes has been adopted by the PCSA bureau of the Construction Industry Manufacturers Association and other U.S. crane manufacturers. In this method, a crane is assigned a nominal capacity rating based on its safe lifting capacity in the direction of least stability (with outriggers set if the crane is so equipped) at a specified operating radius (usually 12 ft) and with a specified boom. A class number consisting of two number symbols follows the nominal rating. The first number indicates the operating radius for the rated capacity. The second number gives the rated load in hundreds of pounds at a 40-ft operating radius using a 50-ft boom. For example, a truck mounted crane having a capacity of 40 tons at a 12-ft operating radius with its basic boom and a capacity of 20,000 lb using a 50-ft boom at a 40-ft operating radius would be classified as a 40-Ton Truck Crane (Class 12-200).

The *tipping load* is that load at which tipping as defined below will occur at a particular operating radius and boom position relative to the carrier. For crawler cranes, tipping is considered to have occurred when any crawler roller is lifted 2 in. away from the treads remaining on the supporting surface. For wheel or truck mounted carriers, tipping is considered to have occurred when all tires on one or more wheels leave the supporting surface.

U.S. Government Occupational Safety and Health Act (OSHA) safety regulations limit the maximum crane load to a percentage of the tipping load as follows:

Type Crane	Safe Load (% of tipping load)
Crawler, no outriggers	75
Crawler, with outriggers set	85
Rubber-tired	85

Other Types of Cranes

Some cranes, such as the wheel mounted crane shown in Figure 2-6, are manufactured for use solely as cranes and do not have the capability of using the front-end attachments shown in Figure 2-2. Another special crane is the tower crane developed in Europe and now widely used in many parts of the world, particularly on high-rise building construction. Tower cranes are available as both mobile cranes and stationary cranes. Some manufacturers also offer tower attachments for their crawler and truck mounted cranes. Because of their wide operating

Figure 2-6 Wheel mounted crane. (Courtesy Bucyrus-Erie Co.)

radius and practically unlimited height capability, tower cranes offer many advantages in multi-story building construction. They also find application in other specialized construction operations such as in prefabrication yards. A mobile tower crane is shown in Figure 2-7.

Climber cranes are designed to support themselves on the structure being erected and to raise themselves as the construction progresses. They are particularly useful in the construction of high-rise buildings and other tall structures such as bridge towers, communication towers, and so on.

Job Planning and Management

As in planning any construction operation, it is essential in planning crane operations to analyze the task to be performed and the working conditions to be expected early in the planning process. For crane operations, the following are some of the principal job factors to be considered:

Figure 2-7 Mobile tower crane. (Courtesy Lorain Division, Koehring Co.)

Type of load

Size and shape of load

Location and required movement of load

Height of lift required

Radius of work permitted by boom and crane selected

Weight of load, including attachments

Clearance for boom and superstructure—give special consideration to electric power lines

OSHA safety regulations specify that no part of a crane or its load may be used closer than 10 ft from a high-voltage line rated 50 kilovolt (kv) or lower. For lines rated over 50 kilovolt, clearance must be 10 ft plus 0.4 in. for each kilovolt over 50 kv. These rules apply unless the power line has been de-energized

HELPFUL HINTS FOR EFFECTIVE CRANE OPERATION

1. Where repetitive lifting is involved, the crane should be positioned for shortest possible swing cycle to reduce cycle time. For heavy lifts, crane should be positioned to lift over end of mounting where it has maximum lift capacity.

2. Crane footing should be checked carefully before lifting capacity or near-capacity loads. Ratings are based on firm, level footings.

3. All overhead obstructions should be inspected carefully before moving a crane or starting lifting operations. Machine should be located so as to avoid any contact with power lines.

4. In attaching loads, a secure hitch must be made and lift started when all helpers are in the clear.

5. Operator should swing crane slowly enough to avoid excessive outward throw of load and over-swinging when machine stops. Crane work is similar to moving a long pendulum which can be controlled only in slow motion. Fast swinging of crane loads will lose more time than it gains through loss of control, and is very dangerous. A tagline device, similar to that used for clamshell buckets, can be attached to loads to control outward swing.

6. Loads should be placed on solid footings so they have no tendency to overbalance when hitch is released.

7. In figuring height of lift, the block, hook, and any sling-slack between hook and load must be included. When making capacity lifts, the entire lifting cycle should be calculated and planned before picking up load. It takes less time and is much safer to check clearances and

position than to lift and try, then reposition and try again. With repetitive lifting, a planned cycle is the best way to high production at low costs.

8. Organize work for minimum travel time. All needed lifts possible in one area should be completed before moving to another location.

9. Booming up and down lengthens the cycle and should be avoided as much as possible on repetitive lifting.

10. With rubber-mounted cranes, outriggers should be securely set before undertaking any near-capacity loads. Footing under jacks must be level and solid.

11. Jerky operations on crane work should be avoided. It is hard on cable and dangerous.

12. Adding a jib to the boom increases the working range both horizontally and vertically, but can reduce lifting capacity.

13. With a given boom length, the steeper the working angle the shorter the working radius. With each degree of boom movement to a more vertical position, there is a corresponding degree of reduction in boom radius—and a corresponding increase in lifting capacity.

14. Level footing avoids swing "up or down hill," requires less power, is faster and safer.

15. When a heavy load is to be lowered from a high position (Example: into a basement or hold of a ship), it is of prime importance that adequate length of hoist cable is assured to facilitate full travel of the block to the lowest point required.

Figure 2-8 Helpful hints for effective crane operations. (Permission to reproduce this material has been granted by the Construction Industry Manufacturers Association (CIMA). CIMA assumes no responsibility for the accuracy of this reproduction.)

and visibly grounded at the point of work or unless insulating barriers not attached to the crane have been erected.

A number of helpful hints for safe and effective crane operations are contained in Figure 2-8 and References 2 and 6.

2-3 CLAMSHELLS

General

The crane-shovel equipped with a crane boom, clamshell bucket, and necessary wire ropes as illustrated in Figure 2-9 is referred to as a *clamshell*. The clamshell is well-suited for such jobs as excavating vertical shafts or footings, moving

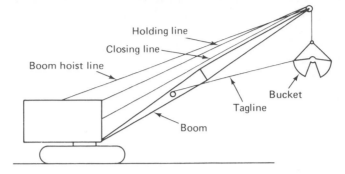

Figure 2-9 Clamshell components.

bulk materials from stockpiles to bins, hoppers, or conveyors, and unloading bulk materials from carriers such as rail cars.

The clamshell bucket consists of two scoops hinged together as illustrated in Figure 2-3. Counterweights are provided to open the bucket when no tension is present on the hoist or closing line (Figure 2-9). When the bucket is open, its weight is carried by the holding line. The tagline is used to prevent the bucket from twisting or swinging during its operation.

Bucket Selection

Clamshell digging action depends primarily on the weight of the bucket. Hence, buckets are provided in a variety of sizes and weights, with removable teeth or without teeth. Buckets are also available with special cutting lips instead of teeth. Heavy buckets are used for medium digging, medium-weight buckets are used for digging loose soils and for general purpose work, and light buckets are used for rehandling loose materials. The orange peel bucket (Figure 2-3) operates on the same principle as the clamshell but is more effective in digging pier holes and shafts and in digging hard materials than is the clamshell.

Since the load on the crane-shovel includes the weight of the empty bucket as well as the weight of the material in the bucket, the lightest bucket capable of efficiently handling the material involved should normally be used. Refer to the manufacturer's "clamshell rating for continuous loading" and tables of lifting capacities at the desired operating radius to obtain the safe load (bucket plus material) for the machine. If a clamshell loading chart is not available, the load should be limited to 80% of the lifting crane capacity chart value for rubber-tired equipment and 90% for crawler mounted equipment to compensate for the shock and load variation encountered in clamshell operation.

Production Estimation

Since tables are not usually available for estimating clamshell production, it is recommended that production estimation be based on previous experience in similar soils with similar machines or that test operations be conducted to determine the cycle time to be used in calculations. Cycle time is the time required for the machine to load, swing, unload, and return to begin the load cycle again. The expected production per hour is then the material moved per cycle multiplied by the number of cycles per hour multiplied by the expected efficiency factor. For the efficiency factor, use the appropriate value from Table 2-6 or use the estimated number of actual working minutes per hour divided by 60. (50 min/hr is widely used as an average value). Thus, the following equation for production estimation is obtained:

$$\text{Production (cy/hr)} = \frac{3{,}600 \text{ (sec/hr)} \times \text{bucket capacity (cy)} \times \text{efficiency factor}}{\text{average cycle time (sec)}}$$

$$(2\text{-}1)$$

Remember that the value found will be expressed in the type of cubic yard that is used for bucket capacity—bank, loose, or compacted.

Example 2-1

PROBLEM: Estimate the production in loose cubic yards per hour for a clamshell given the following information: bucket capacity is $\frac{3}{4}$ LCY; average cycle time is 40 sec; estimated working time per hour is 50 min.

Solution:

$$\text{estimated production} = \frac{3{,}600 \times 0.75 \times 50/60}{40} = 56.25 \text{ or } 56 \text{ LCY/hr.}$$

Job Planning and Management

The weight of material that may be carried in the bucket depends on the bucket weight and machine capacity as previously discussed. The efficient use of a clamshell depends on maintaining an efficient operating cycle. The use of too large a bucket, even if permitted by machine capacity, may increase cycle time and result in decreased production.

Always attempt to position the clamshell so that it is level to avoid swinging uphill or downhill. Uphill swinging requires extra power, is hard on the machine, and usually increases cycle time.

Attempt to set up the job so that the digging radius is the same as the dumping radius. This decreases cycle time by eliminating the need for raising or lowering the boom during each cycle.

2-4 SHOVELS

General

The *shovel*, or crane-shovel equipped with a shovel attachment, was the original member of the crane-shovel family.

The shovel front end includes the shovel boom, dipper stick, dipper (or bucket), shipper shaft, and necessary wire ropes as shown in Figure 2-10. Hydraulic shovels are also available as shown in Figure 2-11.

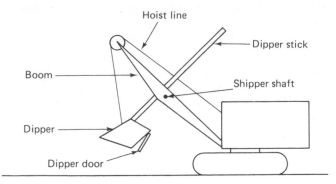

Figure 2-10   Shovel components.

Figure 2-11   Hydraulic shovel. (Courtesy Harnischfeger Corp.)

The shovel digs by a combination of crowd and hoist action illustrated in Figure 2-12. The dipper stick is crowded or retracted by the action of the shipper shaft. After the bucket is filled, the boom is swung laterally to the unloading point and the load dumped by releasing the bucket latch. The ability to force the bucket into the material being dug using crowd action enables the shovel to dig the toughest soils and even soft or fractured rock.

Although the shovel can dig either below or above ground level, it is most efficient when digging from ground level up to about the elevation of the shipper shaft. The "optimum depth of cut" is defined as the vertical distance which allows

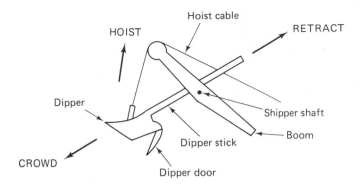

Figure 2-12 Shovel digging action.

the bucket to just fill without undue crowding or hoisting. Maximum production is obtained at the optimum depth of cut. Values of optimum depth of cut vary with the material involved and typical values will be given in the production paragraph of this section.

Although the shovel is capable of digging below ground level, the depth of digging is limited by the length of the dipper stick. Other front-end attachments (such as the hoe or dragline) are better suited than the shovel for excavating below ground level. Since the shovel starts its most efficient digging cycle at ground level, it can form its own roadway as it advances—an important advantage. The shovel is also able to shape the sides of its cut and to dress slopes when required. Material dug by the shovel can be loaded into haul units, dumped onto spoil banks, or side cast into low areas.

Shovel Employment

For efficient digging, the shovel must have a vertical surface to dig against. This surface is known as the *digging face*. A digging face is easily formed when the material to be excavated exists as a bank or hillside, but when the material to be excavated is initially located below ground level, the shovel must dig a ramp down into the material until a digging face of suitable height is created. This process is known as *ramping down*.

Once a suitable digging face has been obtained, the cut is typically developed by using one of the two basic methods of attack (or a variation of these) illustrated in Figure 2-13. The frontal approach allows the most effective digging position of the shovel to be used since the shovel can exert the greatest digging force in this position. This is an important consideration in digging hard materials. Trucks can be loaded on either or both sides of the shovel with a maximum swing usually no greater than 90°. The parallel approach permits fast move-up of the shovel as the digging face advances and it permits a good traffic flow for hauling units. The parallel approach is usually used whenever space is limited and for highway cuts.

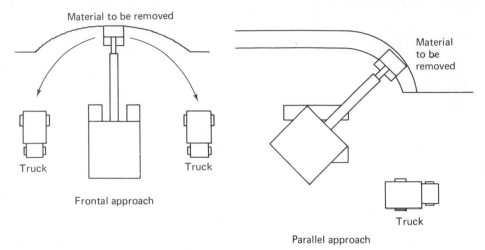

Figure 2-13   Shovel approach methods.

Production Estimation

The PCSA bureau of the Construction Industry Manufacturers Association has made extensive studies of crane-shovel operations and has developed tables used for production planning throughout the industry. Tables 2-3 and 2-4 are based on PCSA data. These tables are intended for diesel-powered, cable-operated shovels unless otherwise indicated. The method of production estimation using the tables is to determine an ideal output (Table 2-3) for the machine and material involved and then to modify this figure by a swing-depth factor (Tables 2-4 and 2-5) and an efficiency factor (using Table 2-6 or the method discussed in Section 2-3).

TABLE 2-3   *Ideal shovel output* (*BCY/hr*)

	Shovel Dipper Size (cu yd)											
Type Material	$\frac{3}{4}$	1	$1\frac{1}{4}$	$1\frac{1}{2}$	$1\frac{3}{4}$	2	$2\frac{1}{2}$	3	$3\frac{1}{2}$	4	$4\frac{1}{2}$	5
Moist loam or sandy clay	165	205	250	285	320	355	405	465	525	580	635	685
Sand and gravel	155	200	230	270	300	330	390	450	505	555	600	645
Common earth	135	175	210	240	270	300	355	405	455	510	560	605
Clay, tough, hard	110	145	180	210	235	265	310	360	405	450	490	530
Rock, well-blasted	95	125	155	180	205	230	275	320	365	410	455	500
Common excavation with rock	80	105	130	155	180	200	245	290	335	380	420	460
Clay, wet, and sticky	70	95	120	145	165	185	230	270	310	345	385	420
Rock, poorly blasted	50	75	95	115	140	160	195	235	270	305	340	375

Note: Above table based on 100% efficiency, 90° swing, optimum depth of cut, material loaded into haul units at grade level. Based on PCSA data.

TABLE 2-4 *Optimum depth of cut for shovels (ft)*

Type Material	Dipper Size (cu yd)											
	$\frac{3}{4}$	1	$1\frac{1}{4}$	$1\frac{1}{2}$	$1\frac{3}{4}$	2	$2\frac{1}{2}$	3	$3\frac{1}{2}$	4	$4\frac{1}{2}$	5
Light, free flowing materials like loam, sand, gravel	5.3	6.0	6.5	7.0	7.4	7.8	8.4	8.8	9.1	9.4	9.7	10.0
Medium materials such as common earth	6.8	7.8	8.5	9.2	9.7	10.2	11.2	12.1	13.0	13.8	14.7	15.5
Harder materials such as tough or sticky clay, soil with rock, or blasted rock	8.0	9.0	9.8	10.7	11.5	12.2	13.3	14.2	15.1	16.0	16.9	17.8

Note: Based on PCSA data.

TABLE 2-5 *Swing-depth factor for shovels*

Depth of Cut (% of optimum)	Angle of Swing						
	45°	60°	75°	90°	120°	150°	180°
40	0.93	0.89	0.85	0.80	0.72	0.65	0.59
60	1.10	1.03	0.96	0.91	0.81	0.73	0.66
80	1.22	1.12	1.04	0.98	0.86	0.77	0.69
100	1.26	1.16	1.07	1.00	0.88	0.79	0.71
120	1.20	1.11	1.03	0.97	0.86	0.77	0.70
140	1.12	1.04	0.97	0.91	0.81	0.73	0.66
160	1.03	0.96	0.90	0.85	0.75	0.67	0.62

Note: Based on PCSA data.

(handwritten annotations: "determine optimum of cut to use", "perhot depth", "Deeper cut", "Shallower cut", "Decrease", "Decrease", "production", "decreasing prod")

TABLE 2-6 *Efficiency factors for construction equipment*

Job Conditions*	Management Conditions			
	Excellent	Good	Fair	Poor
Excellent	0.84	0.81	0.76	0.70
Good	0.78	0.75	0.71	0.65
Fair	0.72	0.69	0.65	0.60
Poor	0.63	0.61	0.57	0.52

*Job conditions are the physical conditions of a job that affect the production rate (not including the type of material involved). These factors include:

    Topography and work dimensions
    Surface and weather conditions
    Specification requirements for work methods or sequence
Management conditions include:
    Skill, training, and motivation of workers
    Selection, operation, and maintenance of equipment
    Planning, job layout, supervision, and coordination of work

Thus,

$$\text{estimated production} = \text{ideal output} \times \text{swing-depth factor} \times \text{efficiency} \quad (2\text{-}2)$$

Note the conditions that apply to the use of Tables 2-3 and 2-6 given at the bottom of the tables. To utilize Table 2-5, divide the actual depth of cut by the optimum depth of cut found in Table 2-4 and express this result as a percentage.

Interpolation in both percent of optimum depth and angle of swing should be used to find values between those given in the table.

Example 2-2

PROBLEM: Given the following information on a shovel excavation job, determine the hourly production in bank cubic yards that can be expected.

Shovel = 3 cu yd.

Swing angle = 75°.

Average depth of cut = 17 ft.

Material = tough clay.

Job conditions = good.

Management = good.

Solution:

Ideal output = 360 BCY/hr (Table 2-3).

Optimum depth of cut = 14.2 ft (Table 2-4).

Actual depth/optimum depth = 17/14.2 = 1.2.

Swing-depth factor = 1.03 (Table 2-5).

Efficiency factor = 0.75 (Table 2-6).

Estimated production = 360 × 1.03 × 0.75 = 278.1 BCY/hr.

Normally, the result should be rounded off to the nearest whole cubic yard (after any volume change calculations required have been performed) since the data used do not support any higher degree of accuracy. If a 50-min work hour had been used instead of the efficiency factor from Table 2-6, then the estimated production in the above situation would have been

$$\text{estimated production} = 360 \times 1.03 \times \frac{50}{60} = 309 \text{ BCY/hr}$$

Thus, we see that the efficiency factor chosen has a significant effect on the value obtained for the estimated production.

Job Management

In planning a shovel excavation operation it is, of course, necessary to select the proper size of machine and bucket for the material and job conditions involved. In addition, there are a number of other factors to be considered. Some of these include the following:

Swing Angle: In order to achieve maximum shovel production, it is necessary to maintain the minimum possible swing angle during excavation and loading (refer to Table 2-5). Thus, the shovel loading and dumping positions must be selected in a manner which yields the minimum angle of swing.

Drainage: Will drainage be a problem, particularly after excavation begins? If so, measures to improve drainage must be taken.

Job Access: Is the job accessible to the shovel and to hauling units? The shovel may have to be transported disassembled and reassembled at the job site. Haul roads may have to be built or improved.

Working Surface Conditions: The shovel should be as level as possible while working. The surface must be kept free of holes and large boulders.

Working Face: The shovel should be kept close to the working face for maximum digging efficiency. Avoid excavating too far beyond the boom point.

Most of these principles also apply to the other excavator members of the crane-shovel family. A number of helpful hints for efficient shovel operation are contained in Figure 2-14.

HINTS FOR EFFICIENT SHOVEL OPERATION

The effective use of a shovel is dependent upon conditions being suited to the machine and the machine being properly maintained. Operating and supervisory personnel ability to coordinate their efforts for starting the excavation, material disposal, and advance move planning contributes to efficiency.

The following points are listed as constructive suggestions for improving operating efficiency:

1. Swing should not be started until dipper clears the material.

2. Shovel should not be spotted too close to working face so the dipper retraction avoids striking the crawlers or boom.

3. Crowding too heavily slows up loading.

4. Dipper should not be placed on the ground and swung sideways to clean up loose material.

5. Boom angle should be changed to suit the loading and digging conditions. Banks should be cut in line with teeth.

6. Each individual cut should be made to handle the material required by the specific job with an eye toward aiding the hauling equipment and preventing lost motion.

7. Trucks and cars should be kept in positions that will prevent excessive swinging.

8. The floor of the cut should be kept smooth, even in borrow pit work. A shovel MUST have an even footing for maximum output.

9. The dipper teeth should be kept sharp. (Extra teeth should be purchased for replacements and interchanged while the worn ones are reconditioned.) Sharp points save cable and power; produce more dipper yardage.

10. The shovel should be moved up often enough to insure complete dipper fill each time and prevent digging with extension of handle too far beyond boom point.

11. In digging hard material the top should be taken off first, then the face worked down.

12. In digging blasted rock the face should be kept as nearly vertical as possible. Loose rocks will fall down away from the machine and not roll under the crawlers.

13. Pitch braces on dipper should be adjusted to cut bank at correct angles.

14. Ends of hoist cable should be reversed after reasonable service.

15. Extra cables should be kept on hand at all times

16. Boom hoist cable should be examined regularly, especially the dead end socket.

17. Operators should watch to see that back of dipper is being filled.

Figure 2-14 Hints for efficient shovel operation. (Permission to reproduce this material has been granted by the Construction Industry Manufacturers Association (CIMA). CIMA assumes no responsibility for the accuracy of this reproduction.)

2-5 DRAGLINES

General

The crane-shovel that is equipped with a crane boom, dragline bucket and accessories, fairlead assembly, and necessary cables is known as a *dragline*. The dragline is an extremely versatile machine capable of digging from above machine level to far below machine level. It can handle material ranging from soft to medium hard. Probably the dragline's greatest advantage over other members of the crane-shovel family is its long reach for both digging and dumping. It also has a high cycle speed that is surpassed only by the shovel.

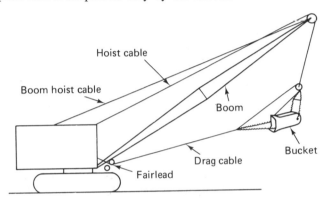

Figure 2-15 Dragline components.

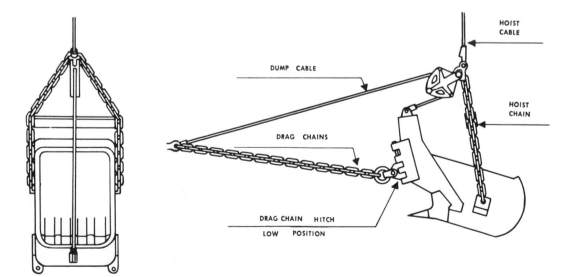

Figure 2-16 Dragline bucket. (Permission to reproduce this material has been granted by the Construction Industry Manufacturers Association (CIMA). CIMA assumes no responsibility for the accuracy of this reproduction.)

Dragline components are shown in Figures 2-15 and 2-16. The hoist cable is used to raise and lower the bucket. Digging is accomplished as the drag cable pulls the bucket through the material to be excavated. The dump cable, through tension on the drag cable, serves to hold up the front of the bucket so that material will not spill where the bucket is being hoisted. Dumping is accomplished by releasing the tension on the drag cable while the bucket is supported by the hoist cable. The length of the dump cable is adjusted so that easy dumping is obtained with minimum spillage for the material being excavated. The fairlead assembly is used to guide the drag cable onto the drum during digging operations.

Dragline buckets are available in a wide range of sizes and types for different applications. Buckets are classified by type as Type I (light duty), Type II (medium duty), and Type III (heavy duty). Both solid (Class S) and perforated (Class P) buckets are available, with or without teeth. Perforated buckets are used when digging underwater or when excavating saturated material because they provide drainage and lighten the load on the machine. However, if it is desired to retain the fines in excavated material, then a perforated bucket should not be used. "Archless" buckets are also available. They eliminate the front cross-member connecting the bucket sides and thus provide easier flow of material into or out of the bucket.

The position at which the drag chain is attached to the bucket (Figure 2-16) serves to change the angle at which the bucket enters the material being excavated. Use of the upper attachment position puts maximum digging pressure on the soil. This position should be used for digging in hard or tough soils.

Dragline Employment

Although the dragline can dig at a level above machine elevation, it is most effective in digging at or below machine level. The dragline does not have the positive digging force of the shovel or hoe but must depend on the weight of the bucket and the angle of attachment of the drag chain to the bucket to provide its digging action. In addition, since the bucket is not rigidly attached to the superstructure, the bucket cannot be held in precise alignment while digging. Therefore, the bucket may bounce, tip, or move sideways when hitting hard material. These actions increase with increased digging depth and with decreasing bucket and machine size. In spite of this, a good dragline operator can maintain a relatively smooth grade while excavating relatively soft materials.

Because of the less precise bucket control, more spillage must be expected in dragline loading operations than in shovel loading. It is recommended that larger haul units be used (compared with shovel loading) to present a larger target for the dragline operator and to minimize spillage. The analysis of combined excavate-haul operations will be discussed in Chapter 4.

The allowable load on a dragline (which includes the weight of the bucket and its load) must be determined from the dragline capacity chart furnished by the machine manufacturer. Do not use the lifting crane capacity chart for this purpose. For dragline use, the power of the machine, length of boom, and weight of material are the major factors governing allowable bucket size.

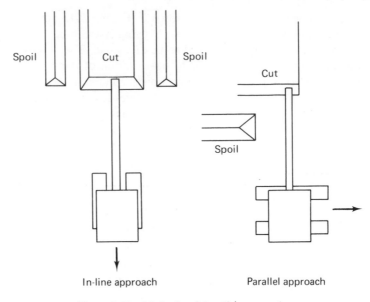

Spoil   Cut   Spoil

Cut

Spoil

In-line approach   Parallel approach

Figure 2-17 Methods of dragline excavation.

The two basic methods of positioning the dragline at the digging face are illustrated in Figure 2-17. For digging trenches and other relatively narrow excavations, the "in-line approach" or "straight-away digging" is usually used. In this approach the machine straddles the center line of the excavation and dumps on one or both sides of the excavation. The machine moves along the center of the excavation line and away from the digging face as excavation progresses. In the parallel approach the machine moves parallel to the digging face as work progresses. The parallel approach is used for wide excavations and for more effective sloping of an embankment.

Dragline Production

Dragline production is calculated in exactly the same manner as is shovel production, except that appropriate draglines tables are substituted. These include

TABLE 2-7 *Ideal dragline output-short boom (BCY/hr)*

Type Material	Bucket Size (cu yd)										
	$\frac{3}{4}$	1	$1\frac{1}{4}$	$1\frac{1}{2}$	$1\frac{3}{4}$	2	$2\frac{1}{2}$	3	$3\frac{1}{2}$	4	5
Light, moist clay or loam	130	160	195	220	245	265	305	350	390	465	540
Sand and gravel	125	155	185	210	235	255	295	340	380	455	530
Common earth	105	135	165	190	210	230	265	305	340	375	445
Tough clay	90	110	135	160	180	195	230	270	305	340	410
Wet, sticky clay	55	75	95	110	130	145	175	210	240	270	330

Note: Based on PCSA data.

dragline ideal output (Table 2-7), optimum depth of cut (Table 2-8), and swing-depth factor (Table 2-9) tables. These tables are based on PCSA data.

TABLE 2-8 *Optimum depth of cut for draglines (ft)*

Type Material	Bucket Size (cu yd)										
	$\frac{3}{4}$	1	$1\frac{1}{4}$	$1\frac{1}{2}$	$1\frac{3}{4}$	2	$2\frac{1}{2}$	3	$3\frac{1}{2}$	4	5
Light, moist clay, loam, sand, and gravel	6.0	6.6	7.0	7.4	7.7	8.0	8.5	9.0	9.5	10.0	11.0
Common earth	7.4	8.0	8.5	9.0	9.5	9.9	10.5	11.0	11.5	12.0	13.0
Tough or wet sticky clay	8.7	9.3	10.0	10.7	11.3	11.8	12.3	12.8	13.3	13.8	14.3

Note: Based on PCSA data.

TABLE 2-9 *Swing-depth factor for draglines*

Depth of Cut (% of optimum)	Angle of Swing							
	30°	45°	60°	75°	90°	120°	150°	180°
20	1.06	0.99	0.94	0.90	0.87	0.81	0.75	0.70
40	1.17	1.08	1.02	0.97	0.93	0.85	0.78	0.72
60	1.25	1.13	1.06	1.01	0.97	0.88	0.80	0.74
80	1.29	1.17	1.09	1.04	0.99	0.90	0.82	0.76
100	1.32	1.19	1.11	1.05	1.00	0.91	0.83	0.77
120	1.29	1.17	1.09	1.03	0.98	0.90	0.82	0.76
140	1.25	1.14	1.06	1.00	0.96	0.88	0.81	0.75
160	1.20	1.10	1.02	0.97	0.93	0.85	0.79	0.73
180	1.15	1.05	0.98	0.94	0.90	0.82	0.76	0.71
200	1.10	1.00	0.94	0.90	0.87	0.79	0.73	0.69

Note: Based on PCSA data.

Example 2-3

PROBLEM: Determine the estimated hourly production in loose cubic yards for the following dragline operation:

Bucket size = $\frac{3}{4}$ cu yd.

Swing angle = 60°.

Average depth of cut = 9 ft.

Material = common earth.

Efficiency = 50 min/hr.

Solution:

Ideal output = 105 BCY/hr (Table 2-7).

Optimum depth of cut = 7.4 ft (Table 2-8).

Actual depth/optimum depth = 9/7.4 = 1.2.

Swing-depth factor = 1.09 (Table 2-9).

Efficiency factor = 50/60 = 0.833.

Volume change factor $= 1.25$ (Table 1-2).

Estimated production $= 105 \times 1.09 \times 0.833 \times 1.25 = 119$ LCY/hr.

Job Management

For maximum efficiency an efficient cycle of dig, hoist, swing, and dump must be obtained. Select the boom length, boom angle, and bucket size and weight (within capacity chart limits) which yield maximum production. Other factors to be considered include adjustment of the dump cable and the attachment position of the drag chains.

HELPFUL HINTS FOR DRAGLINE OPERATION

The foregoing deals specifically with the details of dragline operation. The following summarizes certain practices which will aid in continued efficient operation. The effective use of a dragline is dependent upon conditions being suited to the machine; properly maintaining the machine, and the ability of operating and supervisory personel to coordinate their efforts, starting the excavation, disposing of the material and planning their successive moves in advance. The following points are listed as constructive suggestions for improving operating efficiency.

1. Swing should be started easily when the bucket has cleared the material.

2. The bucket should not be too large for the machine.

    NOTE: The size of the bucket should be judged from the DRAGLINE CAPACITY CHART furnished with the machine, not the lifting crane chart. The efficient use of the dragline is dependent upon an efficient cycle of dig, hoist, swing, and dump. The power of the engine, weight of material, and length of boom are the governing factors for bucket size, not the ability to lift a certain load.

3. In traveling any distance, the bucket should be removed and loaded on a truck.

4. Proper drum laggings and cables should be used.

5. The drag cable guard should be kept in place over the propel gears to protect them from loose material brought in by the drag cable.

6. When changing from crane application to dragline, boom point sheave should be checked to be sure it is large enough in diameter and throat opening for good cable life.

7. The machine should be kept in an efficient position with relation to the work to eliminate digging beyond the point of the boom and unnecessary casting and hoisting.

8. Material should be taken off in layers. In ditch excavation the sides should be taken out ahead of the center, so that the ditch may be kept from narrowing.

9. Pulling the bucket too close to the machine will cause a high roll of material to build up, causing the drag cable to work through it.

10. Bucket teeth should be kept sharp. (Extra teeth should be purchased for replacements and worn ones reconditioned.)

11. Cable and drag chain ends should be changed after a reasonable period of service.

12. Chain hitch on the bucket should be shifted if it fails to dig the material properly.

13. A supply of chain repair links should be kept on hand at all times.

14. Extra cables should be available.

15. Heavy timber mats can be used for work on soft ground. (The mats should be kept level and as clean as possible.)

16. Accessibility to maintenance and operating personnel and hauling equipment should be assured.

Figure 2-18   Helpful hints for dragline operation. (Permission to reproduce this material has been granted by the Construction Industry Manufacturers Association (CIMA). CIMA assumes no responsibility for the accuracy of this reproduction.)

Although the range of the dragline is greater than that of the shovel, the machine should be moved often enough to avoid unnecessary casting and hoisting of the bucket. The most efficient digging area for the dragline bucket is found within an angle of about 15° forward and back from a vertical line through the boom point. Special bucket hitches are available which shorten the drag distance required to obtain a full bucket.

Always consider drainage, access and haul roads, working surface conditions, etc., as for a shovel. When digging in soft soils, low ground pressure tracks or the use of mats may be required.

Excavate in layers when deep cuts are required.

Do not drag the bucket so close to the machine that ridges of material pile up in front of the carrier.

Do not guide the bucket by swinging the superstructure while digging or start swinging before the bucket clears the ground. These actions put undue side stress on the boom and revolving deck.

A number of helpful hints for dragline operations are contained in Figure 2-18.

2-6 HOES

A crane-shovel that is equipped with a hoe front-end attachment is properly called a *hoe*. However, it is frequently referred to as a *backhoe* and sometimes as a *trench hoe, drag shovel, draghoe,* or *pull shovel*. The components of a hydraulic hoe are illustrated in Figure 2-19. Hydraulic machines built primarily for use as hoes are frequently called *hydraulic excavators*.

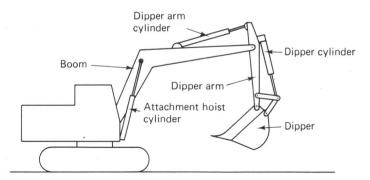

Figure 2-19 Hydraulic hoe components.

Digging is accomplished by pulling the dipper toward the machine. When the dipper is filled, the boom is raised and swung to the unloading position. The load is then dumped by swinging the dipper up and away from the machine.

Cable-operated hoes dig by pulling the dipper toward the machine with a drag cable. A hoist cable is used to extend the dipper as well as to raise or lower the boom. In hydraulic hoes the boom and dipper arms are, raised and lowered by

Figure 2-20 Hydraulic excavator. (Courtesy Bucyrus-Eric Co.)

hydraulic cylinders. In addition, the dipper is pivoted at the end of the dipper arm and controlled by another hydraulic cylinder so that a wrist-like action is provided. Hydraulic hoes may have clamshell, shovel, or loader attachments available. A hydraulic hoe or excavator is shown in Figure 2-20.

Dippers are available in a range of sizes and widths for each machine size. As will be discussed shortly, dipper width is often more important in trenching work than is dipper capacity. Different size side cutters are also available to change the dipper's cutting width.

Since the hoe combines positive digging action with rigid control of the bucket, it is able to dig accurately in all but the hardest materials. The hoe's major advantages are its digging power and its ability to dig below machine level.

Hoe Employment

Although the hoe and dragline can both dig below ground level, they are not really competitors. The dragline has a much wider digging and dumping range, but it possesses less power and accuracy in its digging.

The hoe is especially well-suited for excavating trenches because of the characteristics mentioned above. Since trenches are usually excavated to a specified width and depth, the best measure of production is frequently the number of linear feet of trench excavated to the specified dimensions per unit of time. Therefore, the dipper width chosen (including side cutters when required) should match the required trench width or be only slightly wider. Since there is usually little difference in cycle time for a hoe with various size dippers, the largest dipper available should be used within the machine's capability and the specified ditch width. Production will also be higher with a standard dipper of a specified width than with a smaller dipper having side cutters added to yield the same width of cut.

Hoes, especially hydraulic excavators, are also widely used in laying pipe. Because of the precision with which hydraulic excavators can be controlled, they not only excavate the trench but are also used to pull the man-box, lay bedding, set the pipe into position, and backfill to cover the pipe joints.

Another major application of the hoe is found in making basement excavations and similar precise excavations. Other useful applications of the hoe include cleaning roadside ditches and sloping or grading embankments.

In selecting the proper hoe to be used in a particular application, digging ranges and clearances are important considerations. These include the following:

1. Maximum depth required.

2. Largest working radius required for digging or dumping.

3. Maximum dumping height required.

4. Clearance required for carrier, superstructure, and boom. This includes surrounding buildings or obstacles as well as width of cut required to provide machine clearance in a multi-layer excavation.

Production Estimation

The PCSA has not prepared tables for general hoe production. Hoe production is usually estimated based on dipper capacity and expected machine cycle time and efficiency. Table 2-10 gives typical cycle times for hydraulic excavators.

TABLE 2-10 *Typical cycle times for hydraulic excavators*

Operating Conditions	Cycle Time (sec)	Cycles Per Minute
Fast cycle: small machine	12.0	5.0
	13.3	4.5
Typical range	15.0	4.0
	17.2	3.5
	20.0	3.0
	24.0	2.5
	30.0	2.0
Slow cycle: large machine	40.0	1.5
	60.0	1.0

Job Management

As with other excavators, drainage, access and haul roads, soil conditions, etc., must be considered in job planning.

Although the hoe will dig hard materials, production may be increased and costs reduced by blasting or ripping rock and hardpan prior to excavation. Ledge rock can often be lifted a layer at a time by the dipper once an initial trench has been opened to expose the end of the layers. Do not try to use the dipper as a pick by dropping it at its maximum reach onto hard material.

The hoe can often be used to handle pipe in a pipe laying operation by utilizing a pipe-handling tool attached to the dipper. Use care to insure that the pipe load does not exceed the value in the manufacturer's safe lift capacity chart for the conditions encountered.

A number of helpful hints for effective hoe operation are shown in Figure 2-21.

HELPFUL HINTS FOR EFFECTIVE HOE OPERATION

1. Digging should be planned so that dipper teeth CUT as near as possible to the line of the digging cable.

2. Length and depth of cut should be judged to produce a full dipper at every pass. Full loads on every pass produce more pay dirt than a faster cycle with partly filled dipper. Full loads should be the first objective, followed by speed increases for fast cycles.

3. A hoe will dig fairly hard materials. Where possible, blasting will often be less expensive than bulling through hardpan and rock strata with the hoe dipper.

4. Using the dipper teeth as a pick axe by extending handle to maximum reach, then dropping front end to break ledge rock is very bad practice, the result being serious front end damage.

5. Once the trench is open, ledge rock can be broken by pulling dipper up under the layers. Top layers are pulled first with one or two layers lifted at a time.

Figure 2-21 Helpful hints for effective hoe operation. (Permission to reproduce this material has been granted by the Construction Industry Manufacturers Association (CIMA). CIMA assumes no responsibility for the accuracy of this reproduction.)

This completes the discussion of the excavator members of the crane-shovel family. Scoop loaders and other types of excavators will be covered in Chapter 3.

2-7 PILE DRIVERS

General

As illustrated in Figure 2-3, the most common types of load dropping tools used in conjunction with the crane-shovel are the skull cracker and the pile driver. The skull cracker is a heavy weight that is hoisted by the crane and then swung or allowed to drop free to perform like a huge sledge hammer. It is used to break up pavement, demolish buildings, etc. The simplest form of pile driver, a *drop hammer*, uses a similar action to drive piles. The hammer is hoisted and then dropped onto the pile cap to hammer the pile into the soil.

The basic components of a pile driver attachment are illustrated in Figure 2-22. The pile driver attachment utilizes the crane boom plus adapter plates, leads, catwalk, hammer, a pile cap, and necessary wire ropes. The adapter plates serve to attach the leads to the end of the boom. The leads serve as guides for the drop hammer as it is raised and dropped as well as assist in aligning the pile during

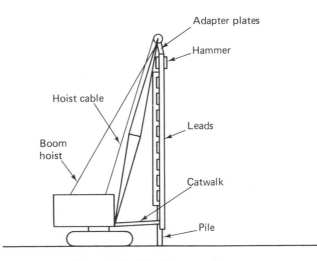

Figure 2-22 Pile driver attachment.

driving. The leads are attached to the foot of the boom by braces which are called a *catwalk*. To reduce energy losses, the hammer is usually made up of two parts: a head and a drop weight. The head is attached to the end of the hoist line and fastens to the drop weight for hoisting of the weight. The head automatically releases the drop weight at the selected height. A pile cap is used to protect the end of the pile from damage by the hammer during driving.

Powered Hammers

For faster pile driving one of several types of powered hammers may be used with the pile driver attachment on the crane-shovel. Powered hammers include hydraulic, steam and compressed air hammers, diesel hammers, and vibratory or sonic hammers. With these hammers the crane is used to support the leads (which provide pile alignment while driving), for hoisting the pile into position, and for holding the hammer in position during driving.

Single-acting steam or compressed air hammers use the force of the fluid to lift the ram (hammer weight) and then allow it to fall freely. Double-acting and differential hammers use fluid pressure to both lift the ram and then to accelerate its fall.

Diesel hammers operate as free-piston or ram-type diesel engines. As the ram falls within its cylinder, it compresses the air below it in the cylinder. Diesel fuel is injected and combustion occurs when the pressure and temperature in the bottom of the cylinder are sufficiently high for ignition. When combustion occurs, it drives the hammer assembly against the pile cap while lifting the ram to begin another cycle.

Vibratory and sonic pile drivers use powered oscillators to produce vibration and vertical forces on the pile. The pile is forced into the soil loosened by vibration under the weight of the pile and the hammer. The sonic hammer uses a high vibration frequency near the resonant frequency of the pile/soil system while vibratory hammers customarily operate at a somewhat lower frequency. A hydraulically powered vibratory driver/extractor is shown in Figure 2-23.

Figure 2-23   Hydraulically powered vibratory driver/extractor. (Courtesy MKT Division, Koehring Co.)

Employment of Pile Drivers

The proper selection and efficient employment of pile drivers are a specialized field which this book will not attempt to cover in detail. The following paragraphs provide only a brief introduction to the topic. For more detailed information, the reader is referred to Reference 1 at the end of this chapter.

The drop hammer provides a rather slow method of driving piles and is best used for driving vertical piles since considerable energy is lost to friction when the hammer leads are not vertical. For efficient driving, the hammer weight should be at least equal to the pile weight, with best results obtained when the hammer weight is about twice as great as the pile weight. The use of a powered hammer and pile leads is recommended for driving piles at an angle (batter piles).

Compressed air, hydraulic, and steam hammers are typically much more effective for pile driving than is the drop hammer. The use of the hydraulic hammer has been growing rapidly because of its quieter manner of operation. The energy of the hammer must be selected to give best results with the size and type of pile being driven. Double-acting and differential hammers may also be effectively used as pile extractors.

Diesel hammers have also become increasingly popular because they are self-powered and are relatively trouble free. However, they cannot be used for pile extraction. They may also fail to operate when a pile is being driven in very soft soil. Vibratory and sonic hammers have proven effective in many soils, especially noncohesive and saturated soils. They may also function as pile extractors.

There have been a number of methods developed to predict pile capacity after driving, including a number of pile driving formulas that utilize driving energy and pile penetration data. Competent geotechnical (soils) engineers and pile driving specialists should select the method most appropriate for a particular case. Frequently, building codes may specify the safe load formula to be used or may require tests (such as a load test) to verify pile capacity.

PROBLEMS

1. If the tipping load of a crawler mounted crane is 40,000 lb in a particular boom position, what is the maximum safe lifting load in this position?
   (a) Without outriggers.
   (b) With outriggers set.

2. What is the minimum clearance required under OSHA safety regulations between a crane and a 60-kv line?

3. How many hours do you estimate would be required to load 400 cu yd of aggregate from a stockpile into trucks using a $\frac{3}{4}$-cu yd clamshell? (Assume bucket capacity = 0.9 LCY.) Average cycle time is estimated at 35 sec. Efficiency is expected to be equivalent to a 50-min hour.

4. If a clamshell rating for continuous loading is not known for a crane being used as a clamshell, what maximum load limit (bucket plus material) would you set when the safe lifting crane load is 30,000 lb?

5. A 3-yd power shovel is being used to load well-blasted rock into trucks. The shovel is operating at optimum depth of cut, swing angle is 120°, job conditions are good, and management is excellent. How many bank cubic yards of rock should be loaded per hour?

6. What would you expect to be the average load per bucket in bank cubic yards for a 3-cu yd shovel loading common earth?

7. A 5-cu yd dragline is excavating sand and gravel from a pit. Average depth of cut is 8 ft, average swing angle is 150°, and efficiency is estimated as a 50-min hour. What is the estimated production in loose cubic yards per hour?

8. Select the most economical dragline from the three listed below to excavate a drainage ditch 10 ft deep having a bottom width of 10 ft and side slopes of 1 : 1. The dragline will move along the center line of the excavation and deposit one-half of the spoil on each side of the cut. Provide 7 ft of clear space between the edge of the cut and the near side of the spoil bank. The soil is common earth with an angle of repose of 37° and swell of 25%. Use an efficiency factor of 0.84. Following are the dragline data:

Size	*Dumping Radius*	*Cost/hour*
1 cu yd	35 ft	$26.25
$1\frac{3}{4}$ cu yd	45 ft	$31.50
2 cu yd	53 ft	$36.75

   (a) What is the estimated cost per bank cubic yard of excavation for the dragline chosen?
   (b) How many machine hours will be required to construct 1 mile of ditch?

9. You are planning to excavate a basement 32 ft by 40 ft to a depth of 5 ft in common earth. A sewer trench 5 ft deep and at least 18 in. wide must also be constructed extending 60 ft from the basement. The hydraulic excavator you have available is equipped with a $\frac{1}{2}$-cu yd bucket 24 in. wide. Estimated cycle time is 15 sec. Efficiency is expected to equal a 50-min hour. Equipment costs are $90 for transporting the excavator to the job and back and $25 per operating hour for the excavator.
   (a) What is the estimated time required to perform the excavation?
   (b) What is the cost per bank cubic yard of excavation?

10. Describe and state the purpose of the following components of a pile driver:
   (a) Leads.
   (b) Catwalk.
   (c) Head.
   (d) Single-acting stream hammer.
   (e) Double-acting hydraulic hammer.

REFERENCES

1. HAVERS, JOHN A., and FRANK W. STUBBS, JR., eds., *Handbook of Heavy Construction*, New York: McGraw-Hill, 1971.

2. *Hydraulic Excavators and Telescoping-boom Cranes*. Milwaukee: PCSA Bureau of Construction Industry Manufacturers Association, 1974.

3. *Technical Bulletin No. 1: Man the Builder*. Milwaukee: PCSA Bureau of Construction Industry Manufacturers Association, 1971.

4. *Technical Bulletin No. 4: Cable-controlled Power Cranes, Draglines, Hoes, Shovels, Clamshells*. Milwaukee: PCSA Bureau of Construction Industry Manufacturers Association, 1954.

5. *TM 5-331B: Lifting, Loading and Hauling Equipment*. Washington, D.C.: U.S. Department of the Army, 1968.

6. *125 Ways to Better Power Shovel-crane Operations*. Milwaukee: PCSA Bureau of Construction Industry Manufacturers Association, 1954.

3

LOADERS
AND OTHER EXCAVATORS

3-1 INTRODUCTION

General

In addition to the members of the crane-shovel family, there are a number of other pieces of equipment suitable for excavating and loading haul units. One of the items of equipment most frequently used for such purposes is illustrated in Figure 3-1. This unit is appropriately called a *wheel loader*. The wheel loader is a wheeled version of a machine also called a *scoop loader*, *front-end loader*, *bucket loader*, or *tractor-shovel*. Track-type loaders are also available as illustrated in Figure 3-2.

Many of these machines, particularly the track-type loader, closely resemble the tractor to be discussed in Chapter 5. The factors affecting speed and usable power (traction, rolling and grade resistance, etc.) are the same for loaders as they are for tractors. These factors will be fully covered in Chapter 5. For the purposes of this chapter, it will be assumed that loader operation takes place on a hard, level surface so that the factors mentioned above may be neglected.

Buckets

Loader bucket capacity is rated somewhat differently from bucket capacity for the members of the crane-shovel family. The nominal or rated bucket capacity

47

Figure 3-1 Articulated wheel loader. (Courtesy Eaton Corp.)

for loaders corresponds to the actual heaped volume capacity [using Society of Automotive Engineers (SAE) standards] of the bucket. This, therefore, expresses bucket volume in loose cubic yards. To convert bucket volume to bank cubic yards for a specific material, two correction factors are applied. First, a bucket efficiency factor or carry factor is applied to change the rated bucket capacity to the average loose volume of material which will be delivered per load. Next, this volume is converted to bank cubic yards by using the load factor explained in Chapter 1. Suggested values for loader bucket efficiency factors are given in Table 3-1.

Loader buckets are available in a wide range of sizes—from less than 1 cu yd to over 20 cu yd. The most common bucket sizes for general purpose use fall in the 2- to 5-cu yd range. Two types of buckets are usually available: a solid or scoop bucket and a multi-segment bucket. The multi-segment bucket has more flexibility than the solid bucket since it can be used like a clamshell, dozer, or scraper as well as a scoop shovel.

In addition to the bucket, other types of attachments available for loaders include hoes, augers, ripper-scarifiers, forklifts, dozer and snow blades, and crane

Figure 3-2 Track loader. (Courtesy Fiat-Allis Construction Machinery, Inc.)

TABLE 3-1 *Suggested bucket efficiency factors for loaders*

Material	Efficiency Factor (%)
Common earth (loam)	100–110
Mixed aggregates	95–100
Uniform aggregates:	
Up to $\frac{1}{8}$ in.	95–100
$\frac{1}{8}$ in. to $\frac{3}{4}$ in.	90–95
Over $\frac{3}{4}$ in.	85–90
Soil with boulders and roots	80–100
Cemented materials	85–95
Rock, blasted	
Well-blasted	80–85
Average	70–80
Poorly blasted	60–70

Figure 3-3.   Backhoe/loader. (Courtesy Bucyrus-Erie Co.)

hooks. Some models of wheel loaders are designed as a combination backhoe and loader. This type of equipment is illustrated in Figure 3-3.

Operating Load

Under SAE standards the maximum operating load of a wheel loader should not exceed 50% of the static tipping load with the loader in full turn position. For track-type loaders, the operating load should not exceed 35% of the static tipping load. Tipping load values may be increased when required by ballasting tires, by using counterweights, or by adding attachments to the rear of the machine. Thus, the machine weight as well as its lifting power determine the size of bucket that may be safely used with the machine.

Employment of Loaders

The loader is a highly versatile piece of equipment. It does not require another item of equipment to support it by leveling, smoothing, or cleaning up its work area since it performs these functions for itself. It has good mobility and can often travel from one construction site to another under its own power. Typical uses of loaders include stripping overburden, stockpiling material, excavating basements, backfilling ditches, loading hoppers and haul units, carrying concrete to forms, and lifting and moving construction materials. When making excavations such as basements, the loader must construct a ramp down into the excavation in the same manner as a shovel. Soils ranging from soft to medium hard may be excavated by loaders.

Selecting a Loader

There are a number of factors involved in selecting a loader for a particular job. Some of these include average and peak production required, type of material to be loaded, clearance required during loading and dumping, capacity and sideboard height of hauling units, etc. Considerations involved in matching loaders with haul units will be covered in Chapter 4.

3-2 WHEEL LOADERS

General

The large rubber tires used on wheel loaders give them excellent job and between-the-job mobility with maximum highway speeds over 25 mph. Their ground pressure is relatively low and may be varied by changing tire size and inflation pressure. However, the vulnerability of its tires to cutting makes it poorly suited to loading sharp shot rock, especially under wet floor conditions, unless its tires are protected in some way (protective mesh chains, etc.). An articulated wheel loader has a frame which is hinged between the front and rear axles, thus providing more maneuverability and a shorter turning radius than a rigid frame. Rigid frame wheel loaders usually steer from the rear axle. Hydraulic steering and control systems are commonly used.

Production Estimation: Rule of 9,000

Estimates of wheel loader production are usually based on bucket capacity and loader cycle time. However, a rule of thumb called the *Rule of 9,000* is sometimes used. This rule states that in stockpile loading a wheel loader operating at 100% efficiency for 2,000 hours per year can produce 9,000 tons per year per flywheel horsepower.

To utilize the Rule of 9,000, it is necessary to apply a job efficiency factor as we have done previously with crane-shovel operations. In addition, if a portion of the loader's time is spent on other operations (such as pit cleanup), a second reduction factor (or job use factor) will be needed. It may even be appropriate to add a third reduction factor to compensate for time lost during spotting (positioning) haul units for the loader. Thus, a loader having a 75% job efficiency factor, 70% job use factor, and a 10% loss of time for truck spotting would yield only

$$\text{production} = 9{,}000 \times 0.75 \times 0.70 \times 0.90 = 4{,}252 \text{ tons/year/hp}$$

Both average and peak loading rates required should be considered when using the Rule of 9,000. After the required loader size (hp) has been determined, then the appropriate bucket size is selected based on the allowable operating load of the loader and the weight of the material being loaded. The size of the bucket must also be appropriate for the hauling unit size as discussed in Chapter 4.

Production Estimation : Cycle Time

The conventional method of estimating loader production is based on average bucket load multiplied by the expected number of cycles per unit of time. This is basically the same method used previously in estimating clamshell and hoe production. Total cycle time is estimated as follows:

$$\text{cycle time} = \text{basic cycle time} + \text{correction factors} + \text{travel time}$$

The basic cycle time includes the time required for loading, dumping, four reversals of direction, and traveling a minimum distance. Field observations indicate that basic cycle times of 0.4 min for articulated loaders under 10-cu yd capacity and 0.5 min for rear steer loaders are average values suitable for production estimation. Travel time (haul and return) is estimated by dividing the travel distance by the appropriate loader speed or by the use of travel time charts such as shown in Figure 3-4.

In order to complete the cycle time estimate, correction factors accounting for the type of material, loading conditions, and operating conditions are added to the

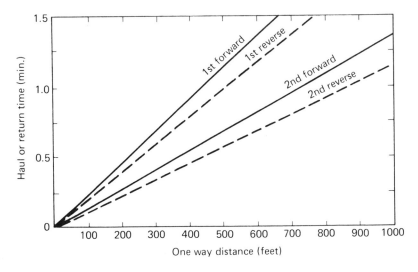

Figure 3-4   Travel time, wheel loader.

basic cycle time and travel time. Suggested values for such correction factors are given in Table 3-2.

TABLE 3-2 *Suggested cycle correction factors for loaders*

Element	Correction Factor (add to basic cycle time)
Material	
Common earth (loam)	+0.01 to +0.03
Mixed aggregates	0.00 to +0.02
Uniform aggregates:	
Up to $\frac{3}{4}$ in.	0.00 to −0.02
$\frac{3}{4}$ in. to 6 in.	0.00
Over 6 in.	+0.03 to +0.16
Soil with boulders and roots	+0.01 to +0.16
Cemented materials	+0.06 to +0.16
Loading Conditions	
Stockpile, under 10 ft high	+0.01
Stockpile, 10 ft or higher	0.00
Dumped by truck	+0.02
Bank	+0.04
Operating Conditions	
Inconsistent operations	+0.02 to +0.04
Small or fragile target	+0.02 to +0.05
Independently owned trucks	+0.02 to +0.04

To complete the production estimate, an average value of bucket load must be computed. This is accomplished by multiplying the rated bucket capacity (which is in loose cubic yards) by the bucket efficiency factor (Table 3-1).

Example 3-1

PROBLEM: Calculate the expected production in loose cubic yards per hour of an articulated loader under the following conditions. Use the travel time chart of Figure 3-4.

Bucket size = 6 cu yd.

Material = $\frac{1}{2}$-in gravel in 15-ft high stockpile.

Haul length = 350 ft (use second gear forward for haul and return).

Dump target = trucks, common ownership, small target.

Operation = consistent.

Job efficiency = 50 min/hr.

Solution:

Bucket efficiency factor = 0.90 (Table 3-1).

Average bucket load = 6 × 0.90 = 5.4 LCY.

Basic cycle time = 0.4 min (articulated loader).

Travel time = 2 × 0.5 = 1.0 min (Figure 3-4).

Correction factors (Table 3-2):
   Material =   −0.01 min
   Loading conditions =   0.00 min
   Operating conditions = +0.03 min (small target)
   Total correction =   +0.02 min

Total cycle time = 0.40 + 1.0 + 0.02 = 1.42 min.

Estimated production = $\dfrac{50}{1.42}$ × 5.4 = 190 LCY/hr.

Figure 3-5   Rough-terrain, high lift forklift. (Courtesy White Construction Equipment Division, White Motor Corp.)

Fork Lifts

Fork lifts are designed primarily for lifting boxes, bundles, or pallets of supplies. Although they are most often used for warehouse operations, fork lifts are also suitable for lifting and loading construction supplies. Fork lift attachments are available for some loaders. However, most fork lifts are especially designed for use as a fork lift. A rough-terrain, high-lift fork lift suitable for construction use is shown in Figure 3-5.

3-3 TRACK-TYPE LOADERS

General

Track-type loaders operate in the same basic way that wheel loaders operate. However, their low ground pressure enables them to operate in areas that would be unsuitable for wheel loaders. Their greater tractive ability makes them suitable for the toughest digging conditions while their steel track resists the cutting action of sharp rock. They can travel on side slopes up to 35% while wheel loaders are limited to about 15%. Their climbing ability enables them to climb 60% grades compared to 30% grades for wheel loaders. Because of their lower travel speed, however, their mobility is much less than that of wheel loaders. Their production will similarly be lower than that of a wheel loader if the haul distance is appreciable.

Production Estimation

Track loader cycle time and production are estimated in exactly the same manner as that used for wheel loaders. However, an average value of 0.35 min for the basic cycle time is suggested. Travel time is determined from an appropriate travel time chart. A typical time chart for track loaders is illustrated in Figure 3-6. Correction factors for material, loading conditions, and operating conditions are the same as for wheel loaders (Table 3-2).

Example 3-2

PROBLEM: Estimate the probable production of a track loader under the following conditions:

Bucket size = 2 cu yd.

Material = mixed aggregates, 20-ft high stockpile.

Haul distance = 125 ft one way (use third gear forward, both directions).

Operation = consistent.

Job efficiency = 0.75.

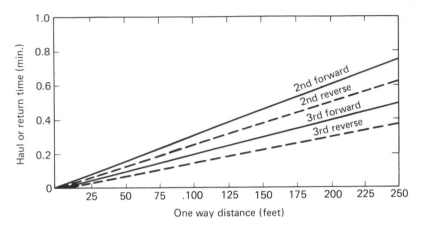

Figure 3-6 Travel time, track loader.

Solution:

Bucket efficiency factor = 0.95 (Table 3-1).

Average bucket load = 2 × 0.95 = 1.90 LCY.

Basic cycle = 0.35 min.

Travel time = 2 × 0.25 min (Figure 3-4).

Correction factors (Table 3-2):

Material =	+0.01 min
Load conditions =	0.00 min
Operating conditions =	0.00 min
Total correction =	+0.01 min

Total cycle time = 0.35 + 0.50 + 0.01 = 0.86 min.

Estimated production = $\dfrac{60}{0.86}$ × 1.9 × 0.75 = 99 LCY/hr.

Job Management

The factors to be considered in job management of loaders are similar to those previously discussed for excavators of the crane-shovel family. Considerations include drainage, job access, working surface conditions, and conditions at the working face. Optimum positioning of the loader and haul units is necessary in order to minimize loading, maneuver, and dump times. When handling sticky material, a multi-segment bucket is usually more efficient than is a solid bucket. Production in loading hard material may often be increased by blasting or ripping prior to excavating.

3-4 OTHER EXCAVATORS AND LOADERS

General

In addition to the members of the crane-shovel family and front-end loaders, there are several other types of equipment used for excavating and loading. These include dredges, scrapers (which will be covered in Chapter 6), elevating loaders, bucket wheel excavators, and trenching machines. All of these machines except dredges and scrapers combine some type of cutter with a conveyor belt for moving excavated material to hauling units. Dual belts are frequently used with these machines so that haul units may be loaded on alternate sides of the machine, thus eliminating the time lost in spotting haul units.

Elevating Loaders

Elevating loaders combine a scoop-type cutting head with a belt loader. Usually, the machine is towed by one or more earthmoving tractors, although the loader may be equipped with its own propelling units. Such machines are capable of excavating and loading several thousand bank cubic yards of material per hour under favorable conditions.

Bucket Wheel Excavators

Bucket wheel or continuous-wheel excavators combine large bucket wheel-type cutters with a belt loading system. These machines are usually self-propelled and designed for high production rates. They, like the elevating loader, are best suited to large, relatively flat excavation areas. These machines may yield low excavating and loading unit costs on large earthmoving projects such as dam construction, mining, etc.

Trenching Machines

Trenching machines were developed to dig trenches for cables and pipelines faster and more economically than excavators like the hoe or the dragline. Major types of trenching machines include the continuous chain trencher, the ladder-type trencher, and the wheel-type trencher.

Continuous-chain trenchers use a series of digging scoops spaced along a continuous, power-driven chain to provide digging action. They are used for relatively shallow, narrow trenches and are available in very small sizes for laying power and communications lines. They are usually limited in digging depth to 6 ft or so. A wheeled trencher equipped with hoe and blade is shown in Figure 3-7.

Figure 3-7   Wheeled trencher. (Courtesy Parsons Division, Koehring Co.)

Ladder-type trenchers operate somewhat like continuous-chain trenchers but are larger and heavier. They are available in a wide range of bucket widths and can excavate to a depth of 30 ft or more. Thus, they have the greatest depth capability of any trencher.

Wheel-type trenchers use a bucket wheel-type of excavator for digging action. They are also available in a range of bucket widths but are limited in digging depths to around 10 ft.

Dredges

Although they are primarily used for underwater excavation, dredges are a form of excavating and loading equipment. Dredges may be used for general excavation when adequate water is available to permit complete saturation of material during excavation and transportation. Transportation of excavated material is usually accomplished by pumping the material in a slurry form through a pipeline to a dumping site. Because of environmental considerations, care must be taken to insure that waters outside the construction area are not polluted by the excavation or dumping action.

PROBLEMS

1. Estimate the production in loose cubic yards per hour for a 5-cu yd track loader engaged in basement excavation. Material is common earth loaded into trucks on the excavation floor.

2. How many hours should it take a 5-cu yd wheel loader to load a 2,000-cu yd stockpile (8 ft high) of 1-in. gravel into rail cars if the average haul distance is 600 ft one way? Use the travel time curve of Figure 3-4, articulated loader. Use second gear forward for hauling and returning. Assume a 50-min hour.

3. Use the Rule of 9,000 to find the size loader needed to move 1 million tons per year (2,000 hr of operation) if the job efficiency is 80% and the job use factor is 100%.

4. How many bucket loads would be required for a 5-yd loader to fill a dump truck holding 16 LCY with mixed aggregates?

5. Would you recommend a wheel loader or track loader in the following situations? Why?
   (a) Operating on a side slope of 20%.
   (b) Operating up a grade of 40%.
   (c) On a level haul of 400 ft.

REFERENCES

1. *Basic Estimating* (3rd ed.). Melrose Park, Illinois: Construction Equipment Division, International Harvester Company, n.d.

2. *Caterpillar Performance Handbook*. Peoria, Illinois: Caterpillar Tractor Co., 1975.

3. HAVERS, JOHN A., and FRANK W. STUBBS, JR., eds., *Handbook of Heavy Construction*, New York: McGraw-Hill, 1971.

4. "Match Your Loaders With Your Trucks," *Construction Equipment*, Vol. 23, No. 12 (December 1972), pp. 46–48.

5. *Production and Cost Estimating of Material Movement with Earthmoving Equipment.* Hudson, Ohio: TEREX Division, General Motors Corporation, 1970.

4

TRANSPORTATION
OF EXCAVATION

4-1 INTRODUCTION

General

The transportation of excavation (or hauling) is one of the biggest jobs done in earthmoving. A wide range of equipment is available for transporting excavated material, including dump trucks, wagons, scrapers, conveyor belts, and trains. Although scrapers may be used solely as a transportation device, they are designed to both load and haul excavation. Since the operation of scrapers will be covered in detail in Chapter 6, scrapers will not be further discussed here.

Belt-type conveyors are another form of transportation equipment. Most conveyors used in construction work are either portable units used for the movement of materials within a limited area or they are components of another piece of equipment, such as the elevating loader mentioned in the previous chapter. However, large custom-built conveyor installations provide efficient transportation when there is sufficient material involved to justify their large capital cost. They have been used economically on several large dam projects.

Trains are also sometimes used to transport excavation. Conventional freight trains may be used if the hauling distance is great and tracks are located near the loading and dumping areas. In most cases, however, specially constructed narrow gage railroads are used in hauling excavation. These are frequently used to remove the spoil from tunneling operations. They have also been used for large earth-moving projects such as dams. High initial costs make their economic characteristics similar to those of conveyors.

Trucks and Wagons

In spite of the availability of the equipment mentioned above, the vast majority of excavation hauling is accomplished by trucks and wagons. The heavy-duty, rear-dump truck is the piece of equipment most used. The dump truck permits great flexibility in hauling operations, and the highway (or over-the-road) model can be moved rapidly between job sites. Dump trucks are available in a wide variety of sizes and types. For example, they are available with standard or rock-type bodies, with diesel or gas engines, with either 2-wheel or 4-wheel drive, with two or three axles, and they can be designed for either highway or off-highway use, etc. Dump trucks operated over public highways are usually limited to 8 ft or less in maximum width and have gross weight and axle load limitations. Trucks designed for off-highway use, however, can be larger and heavier with maximum payloads of several hundred tons. The use of such large off-highway units is increasing in large earthmoving projects. An 85-ton off-highway truck having a heaped capacity of 67 cu yd is shown in Figure 4-1.

Figure 4-1 85-ton off-highway truck. (Courtesy Caterpillar Tractor Co.)

Wagons are earthmoving trailers pulled by tractors or truck-tractors. They are sometimes referred to as *pure haulers* because they have many characteristics of tractor-scrapers, but they are designed for hauling only. They are available in bottom-dump, end-dump, and side-dump models. Although wagons are independent pieces of equipment, some are especially designed to work with a particular make and model of tractor.

A 70-ton, bottom-dump wagon equipped with two longitudinal flow gates is shown in Figure 4-2. Longitudinal flow dumping is desirable for windrowing and stockpiling. Cross flow dumping permits better spreading of base materials and aggregates. A wagon capable of either cross flow or longitudinal flow spreading is shown in Figure 4-3. A 22-ton articulated rear-dump rock hauler is shown in Figure 4-4.

Figure 4-2 70-ton bottom-dump wagon. (Courtesy Challenge-Cook Brothers, Inc.)

Payload

The allowable hauling load or payload, of a dump truck or wagon is usually expressed in three ways: (1) the rated capacity in pounds; (2) its struck capacity in cubic yards, and (3) its heaped capacity (usually at a 2 to 1 slope) in cubic yards. Since an overload will result in reduced performance as well as increased mechanical wear on the unit, the rated capacity in pounds should not be exceeded. Because hauling normally involves soil in the loose state, the maximum load that can be carried is thus limited by the smaller of (1) the heaped volume capacity or (2) the

Figure 4-3 Switch gate bottom-dump wagon. (Courtesy CMI Corp.)

Figure 4-4 Articulated rock hauler. (Courtesy ISCO Manufacturing Co.)

rated weight capacity. If the weight of the heaped volume of loose soil is less than the allowable weight load, the body capacity may be increased by the use of sideboards. The struck volume capacity of a hauler is not usually a governing limit but may be limiting in the case of hauling fluid materials such as concrete or when limitations on spillage are in effect.

4-2 DETERMINING THE NUMBER OF HAUL UNITS REQUIRED

The Haul Cycle

In order to determine the number of haul units required to service an excavator, it is necessary to compute the time required for a haul unit to make one complete cycle. Total cycle time is found by summing the time required for each of the following components of the haul cycle:

1. Load—at the excavator/loader
2. Haul—from the loader to the unloading site
3. Dump—at the unloading site, including maneuvering
4. Return—travel back to the loader
5. Spot—move into loading position at the loader

The components of the haul cycle are often grouped together into fixed time and variable time as expressed in Equation 4-1.

$$\text{haul cycle} = \text{fixed time} + \text{variable time} \qquad (4\text{-}1)$$

Fixed time is the sum of load, dump, and spot times. These times are referred to as fixed since they do not depend on the haul distance or travel speed of the haul unit. Fixed times can usually be closely estimated for a particular operation.

The components of travel time are added together and referred to as *variable time*. Travel time can be found by using travel time curves or by dividing the travel distance by the average speed of the haul unit. The use of travel time curves and the method of computing vehicle speed from performance curves will be covered in the following chapter. In this chapter it will be assumed that travel time for the haul unit is already known.

Computing Loading Time

The time required to load a haul unit may be estimated by applying the appropriate one of the following:

$$\text{load time} = \frac{\text{haul unit capacity}}{\text{loader production at } 100\% \text{ efficiency}} \qquad (4\text{-}2)$$

$$\text{load time} = \text{number of bucket loads} \times \text{excavator cycle time} \qquad (4\text{-}3)$$

In Equation 4-2 the loading rate at 100% loader efficiency is used because a loader typically operates at or near 100% efficiency when actually engaged in loading. Either bank cubic yards or loose cubic yards may be used in Equation 4-2 as long as the same unit is used in both the numerator and denominator.

Calculating the Number of Haul Units Required

The traditional (or rational) method of calculating the number of haul units required to service a loader involves providing a sufficient number of haul units so that theoretically there is always a haul unit available for loading. Note that this is a deterministic method which assumes that actual loading and travel times will require exactly the length of time used in the formula. As you know, in the real world these operations are subject to some variance. Thus, at times there will actually be no truck available at the loader or there will exist a waiting line at the loader. The application of a mathematical technique called *queueing theory* takes into account the variability of loading and travel times and thus provides a more realistic estimate of actual system production. The use of queueing theory in loading and hauling operations will be covered in the following section.

In the traditional method the number of haul units required to service a loader is given by Equation 4-4.

$$\frac{\text{number of units required}}{(N)} = \frac{\text{total haul unit cycle time}}{\text{load time}} \tag{4-4}$$

In general, the result should be rounded up to the next integer to insure that at least the number of trucks theoretically required to service the loader is provided. The effect on production when fewer than N trucks are used will be discussed shortly.

The production of the loading and hauling system using this procedure is assumed to be the same as the normal production of the loader itself. The normal production of the loader, of course, includes a job efficiency factor as used in previous chapters. To review the procedure, system output is assumed to equal loader output applying the usual job efficiency factor. However, haul units are provided on the basis of a loading time calculated at 100% loader efficiency. The following examples will illustrate the use of the procedure.

Example 4-1

PROBLEM: Using the shovel conditions given in Example 2-2 and a truck travel time of 0.5 hr, determine how many trucks having a capacity of 20 BCY would be required to fully service the shovel. How many bank cubic yards per hour will be produced by this combination?

Solution:

Load time $= \dfrac{20}{360 \times 1.03} = 0.054$ hr (Equation 4-2).

Travel time $= 0.5$ hr (given).

Number of units required $= \dfrac{0.054 + 0.5}{0.054} = \dfrac{0.554}{0.054} = 10.3$ ∴ Use 11 trucks.

(Equation 4-4).

Expected production $= 278$ BCY/hr (from Example 2-2).

Example 4-2

PROBLEM: If the 3-cu yd shovel referred to in Examples 2-2 and 4-1 had a dipper cycle time of 30 sec, how many trucks would be required to service this shovel? The soil is tough clay with a bucket fill factor of 0.65. Assume that the trucks will still carry 20 BCY of soil.

Solution:

Average bucket capacity $= 3 \times 0.65 = 1.95$ BCY.

Number of dippers required to fill haul unit $= \dfrac{20}{1.95} = 10.3$ (use 11 dippers).

Load time $= \dfrac{11 \times 30}{3,600} = 0.092$ hr (Equation 4-3).

Number of units required $= \dfrac{0.092 + 0.5}{0.092} = \dfrac{0.592}{0.092} = 6.43$ (use 7 trucks)

(Equation 4-4).

Note that the use of 11 shovel cycles to fill a truck will require that each bucket load be slightly reduced or that the last bucket carry only $\frac{3}{10}$ of a full load. In either case, spillage during loading must be minimized.

Effect of Reduced Haul Units

If the rational method of determining the number of haul units required is used, no increase in production occurs if more than the required number of haul units is provided. However, if less than the required number is available, the expected production is reduced in proportion to the shortage. This is expressed by

$$\begin{array}{c} \text{expected production} \\ (\text{units} < N_{\text{reqd}}) \end{array} = \frac{\text{actual number of units}}{\text{required number of units}} \times P_{\text{nor}} \qquad (4\text{-}5)$$

Example 4-3

PROBLEM: In the situation of Example 4-1, only 8 trucks are available. What is the expected production of this system?

Solution:

Expected production $= \dfrac{8}{10.3} \times 278 = 216$ BCY/hr (Equation 4-5).

Cost Analysis

In the method described above the number of trucks of a given size required to service a specific loader does not consider economic factors. In reality, cost performance (or cost per unit of production) is usually a prime consideration in planning an excavation and hauling operation. Such an analysis may be made by using the production estimate obtained by the above methods. Cost performance is calculated by using the following:

$$\frac{\text{cost performance}}{\text{(haul only)}} = \frac{\text{cost of haul fleet/unit time}}{\text{production/unit time}} \tag{4-6}$$

$$\frac{\text{cost performance}}{\text{(load and haul)}} = \frac{\text{total cost of equipment fleet/unit time}}{\text{production/unit time}} \tag{4-7}$$

By considering a range of haul unit sizes and by varying the number of haul units, we may find the size and number of haul units to service a specific loader that yield the lowest cost per unit of production. This is a trial-and-error procedure that may become time-consuming if done manually. In the queueing theory method to be presented in the next section, cost performance is used directly to select the optimum number of trucks of a particular size to service a specific loader. Simulation methods to be discussed in Chapter 14 may be used to more rapidly optimize the design of an excavation and hauling system.

Example 4-4

PROBLEM: If the shovel conditions described in Example 2-2 and if the following information are used, what size truck and how many of them should be used to provide the lowest loading and hauling cost? What will be the production of this system? Shovel cost is $36.00/hr. Truck data are as follows:

Size Truck (BCY)	Cost ($/hr)	Travel Time (hr)
14	13.00	0.46
18	15.00	0.48

Solution:

$$\text{Load time (14 cy)} = \frac{14}{360 \times 1.03} = 0.0378 \text{ hr.}$$

$$\text{Load time (18 cy)} = \frac{18}{360 \times 1.03} = 0.0485 \text{ hr.}$$

$$N \text{ (14 cy)} = \frac{0.0378 + 0.46}{0.0378} = \frac{0.4978}{0.0378} = 13.2.$$

$$N \text{ (18 cy)} = \frac{0.0485 + 0.48}{0.0485} = \frac{0.5285}{0.0485} = 10.9.$$

Size	Number	Production (BCY/hr)	Fleet Cost ($/hr)	Cost Performance ($/BCY)
14	14	278	218	0.784
14	13	274	205	0.748
18	11	278	201	0.723
18	10	255	186	0.729

Optimum solution = 11 18-cu yd trucks @ $0.723/BCY.

Expected production = 278 BCY/hr.

There are, of course, other considerations involved in selecting the number and type of haul units to be used for a particular operation. These will be discussed later in this chapter.

4-3 QUEUEING THEORY METHOD

Introduction

In 1947 Palm presented a method based on the theory of finite queues to determine the optimum distribution of repairmen servicing automatic machines. Variations of this model (often called the *Swedish Machine Model*) have been successfully applied to a number of similar problems. Spaugh is credited with first applying the model to a construction excavation and hauling operation (see Reference 3 at the end of this chapter).

Although queueing theory is founded on a somewhat complex mathematical basis, it may be easily applied to the excavation and hauling problem. The mathematical derivation of the procedure will not be covered here. Interested readers are referred to Reference 3 at the end of this chapter for information on model derivation. However, it should be recognized that the procedure assumes a specific distribution (a Poisson or exponential distribution) for both travel time and loading time. Field observations indicate that such a distribution is not precisely correct for the earthmoving situation. In spite of this, actual production is close to that predicted by the model. These factors will be discussed further later in this section.

Application of Queueing Theory

The following terms and symbols will be used in applying queueing theory to the loading and hauling problem:

n = number of haul units in the fleet.

a = mean arrival rate of a particular haul unit (arrivals/hr).

l = mean loading rate of the excavator (units/hr).

r = ratio of arrival rate to loading rate.

P_0 = probability that no haul unit is available at the loader.

P_t = probability that one or more haul units are available at the loader.

Since there must be either a haul unit at the loader or no haul unit at the loader, the sum of P_t and P_0 must equal one. Hence,

$$P_t = 1 - P_0 \qquad (4\text{-}8)$$

In order to find P_t and P_0, it will be necessary to calculate the ratio r defined above. A simple equation for computing r can be developed as follows:

$$a = \text{arrival rate} = 1 \div \text{travel time}$$

$$l = \text{loading rate} = 1 \div \text{loading time} = 1 \div \frac{\text{truck capacity}}{\text{loader production}}$$

$$r = \frac{\text{loading time}}{\text{travel time}} = \frac{\text{truck capacity}}{\text{loader production} \times \text{travel time}} \qquad (4\text{-}9)$$

In Equation 4-9 loader production is based on 100% loader efficiency for the reasons discussed in the previous section.

The value of P_0 (probability of having no haul unit available for loading at any particular instant) and, thus, the value of P_t will depend on the number of haul units in the fleet as well as the ratio r. Tables for determining the value of P_t have been developed. Linear interpolation may be used for values of r between those given in Table 4-1. For greater accuracy, more detailed tables may be used or the value of P_0 may be calculated from the following equation:

$$P_0 = \left[\sum_{i=0}^{n} \frac{n!}{(n-i)!} (r)^i \right]^{-1} \qquad (4\text{-}10)$$

Although this equation looks complex, it is easily computed by using a hand calculator, as the following example will show.

Example 4-5

PROBLEM: Use the shovel and truck of Example 4-1 with a truck travel time of 0.54 hr and a fleet of 5 trucks. What is the probability that there will be a truck available for loading at any particular instant?

Solution:

$$r = \frac{20}{360 \times 1.03 \times 0.54} = 0.1 \text{ (Equation 4-9)}.$$

$$P_0 = \left[\sum_{i=0}^{5} \frac{5!}{(5-i)!} (0.1)^i \right]^{-1} \text{ (Equation 4-10)}.$$

$$P_0 = \left[\frac{5!}{5!}(0.1)^0 + \frac{5!}{4!}(0.1)^1 + \frac{5!}{3!}(0.1)^2 + \frac{5!}{2!}(0.1)^3 + \frac{5!}{1!}(0.1)^4 \right.$$
$$\left. + \frac{5!}{0!(=1)}(0.1)^5 \right]^{-1}.$$

$$P_0 = [1 + (5)(0.1)^1 + (5)(4)(0.1)^2 + (5)(4)(3)(0.1)^3 + (5)(4)(3)(2)(0.1)^4$$
$$+ (5)(4)(3)(2)(1)(0.1)^5]^{-1}.$$

TABLE 4-1 *Probability of haul unit being available (P_t)*
Number of Haul Units

r	3	4	5	6	7	8	9	10	11	12	13	14	15
.01	.030	.040	.049	.059	.069	.079	.089	.099	.109	.119	.129	.138	.148
.02	.059	.078	.098	.117	.137	.156	.176	.195	.215	.234	.253	.274	.292
.03	.087	.116	.145	.174	.203	.231	.260	.288	.317	.345	.373	.401	.429
.04	.115	.153	.191	.229	.266	.304	.341	.378	.414	.450	.486	.522	.556
.05	.142	.189	.236	.282	.328	.373	.418	.462	.506	.548	.590	.631	.670
.06	.169	.224	.279	.333	.386	.439	.490	.541	.590	.637	.682	.726	.766
.07	.194	.258	.320	.382	.442	.501	.558	.613	.665	.715	.762	.804	.843
.08	.220	.291	.361	.429	.495	.559	.620	.678	.732	.782	.827	.866	.900
.09	.244	.323	.399	.473	.545	.613	.676	.736	.789	.837	.876	.911	.938
.10	.268	.353	.436	.515	.591	.662	.727	.785	.837	.880	.916	.943	.964
.11	.291	.383	.471	.555	.634	.706	.771	.828	.875	.914	.943	.964	.979
.12	.314	.412	.505	.593	.673	.746	.810	.863	.906	.939	.962	.978	.988
.13	.335	.439	.537	.627	.709	.782	.843	.892	.930	.957	.975	.987	.993
.14	.357	.465	.567	.660	.742	.813	.871	.915	.948	.970	.984	.992	.996
.15	.377	.491	.596	.690	.772	.840	.894	.934	.962	.979	.989	.995	.998
.16	.397	.515	.622	.718	.799	.864	.914	.949	.972	.986	.993	.997	.999
.17	.416	.538	.648	.743	.823	.885	.930	.960	.979	.990	.996	.998	
.18	.435	.560	.672	.767	.844	.902	.943	.969	.985	.993	.997	.999	
.19	.453	.581	.694	.788	.863	.917	.954	.976	.989	.995	.998		
.20	.470	.602	.715	.808	.879	.930	.963	.982	.992	.997	.999		
.21	.487	.621	.735	.826	.894	.941	.970	.986	.994	.998			
.22	.504	.639	.753	.842	.907	.950	.975	.989	.995	.998			
.23	.519	.657	.770	.857	.919	.958	.980	.991	.996	.999			
.24	.535	.673	.786	.871	.929	.964	.984	.993	.997				
.25	.549	.689	.801	.883	.937	.970	.987	.995	.998				
.26	.564	.704	.815	.894	.945	.974	.989	.996	.999				
.27	.577	.719	.828	.904	.952	.978	.991	.997					
.28	.591	.732	.839	.913	.957	.981	.993	.997					
.29	.603	.745	.851	.921	.962	.984	.994	.998					
.30	.616	.757	.861	.928	.967	.986	.995	.998					
.31	.628	.769	.870	.935	.971	.988	.996	.999					
.32	.639	.780	.879	.941	.974	.990	.997						
.33	.650	.791	.887	.946	.977	.991	.997						
.34	.661	.800	.895	.951	.980	.993	.998						
.35	.671	.810	.902	.955	.982	.994	.998						
.36	.681	.819	.909	.959	.984	.995	.998						
.37	.691	.827	.915	.963	.986	.995	.999						
.38	.700	.835	.920	.966	.987	.996							
.39	.709	.843	.925	.969	.989	.996							
.40	.718	.850	.930	.972	.990	.997							
.41	.726	.857	.935	.974	.991	.997							
.42	.734	.863	.939	.976	.992	.998							
.43	.742	.870	.943	.978	.993	.998							
.44	.750	.875	.946	.980	.994	.998							
.45	.757	.881	.950	.982	.994	.998							
.46	.764	.886	.953	.983	.995	.999							
.47	.771	.891	.956	.985	.995								
.48	.777	.896	.958	.986	.996								
.49	.783	.900	.961	.987	.996	.999	.999	.999	.999	.999	.999	.999	.999

TABLE 4-1 *(cont.)*

r	3	4	5	6	7	8	9	10	11	12	13	14	15
.50	.780	.905	.963	.988	.997	.999	.999	.999	.999	.999	.999	.999	.999
.51	.795	.909	.965	.989	.997								
.52	.801	.913	.968	.990	.997								
.53	.807	.916	.969	.990	.997								
.54	.812	.920	.971	.991	.998								
.55	.817	.923	.973	.992	.998								
.56	.822	.926	.974	.992	.998								
.57	.827	.929	.976	.993	.998								
.58	.831	.932	.977	.993	.998								
.59	.836	.935	.978	.994	.999								
.60	.840	.938	.980	.994									
.61	.844	.940	.981	.995									
.62	.848	.942	.982	.995									
.63	.852	.945	.983	.995									
.64	.856	.947	.984	.996									
.65	.860	.949	.985	.996									
.66	.863	.951	.985	.996									
.67	.867	.953	.986	.997									
.68	.870	.954	.987	.997									
.69	.873	.956	.987	.997									
.70	.877	.958	.988	.997									
.71	.880	.959	.989	.997									
.72	.882	.961	.989	.998									
.73	.885	.962	.990	.998									
.74	.888	.964	.990	.998									
.75	.891	.965	.991	.998									
.76	.893	.966	.991	.998									
.77	.896	.967	.992	.998									
.78	.898	.968	.992	.998									
.79	.901	.970	.992	.998									
.80	.903	.971	.993	.998									
.81	.905	.972	.993	.999									
.82	.907	.973	.993										
.83	.910	.973	.994										
.84	.912	.974	.994										
.85	.914	.975	.994										
.86	.915	.976	.994										
.87	.917	.977	.995										
.88	.919	.978	.995										
.89	.921	.978	.995										
.90	.923	.979	.995										
.91	.924	.980	.996										
.92	.926	.980	.996										
.93	.928	.981	.996										
.94	.929	.981	.996										
.95	.931	.982	.996										
.96	.932	.983	.996										
.97	.933	.983	.997										
.98	.935	.984	.997										
.99	.936	.984	.997	.999	.999	.999	.999	.999	.999	.999	.999	.999	.999

$$P_0 = [1 + 0.500 + 0.200 + 0.060 + 0.012 + 0.001]^{-1}.$$

$$P_0 = [1.773]^{-1} = \frac{1}{1.773} = 0.564.$$

$$P_t = 1 - P_0 = 1 - 0.564 = 0.436 \text{ (Equation 4-8)}.$$

Optimum Number of Haul Units

The expected production of an excavation/haul system using queueing theory is determined by multiplying the normal production of the excavator by the probability of having a haul unit available at any instant. Thus, expected production is the normal production of the excavator multiplied by P_t (for the number of haul units being used). In the previous example the expected production of the system would be only 43.6% of the normal excavator production.

$$\text{expected production} = \text{normal excavator production} \times P_t \qquad (4\text{-}11)$$

When queueing theory is used, the optimum number of haul units for a particular operation is selected as the combination of excavator and haul units which yields the best cost performance (i.e., lowest unit cost of production). Cost performance over a range of haul unit numbers is determined and the optimum number of units selected. An approximate value of the optimum n may be found by taking the reciprocal of r. A range of n values about this value should then be investigated. The following example will illustrate the procedure.

Example 4-6

PROBLEM: Use the truck and shovel data of Example 4-5 with a shovel cost of $36/hour and a cost per truck of $16/hour. What is the optimum number of haul units to use for this operation?

Solution:

$r = 0.1$ (from Example 4-5).

Normal production of shovel $= 278$ BCY/hr (from Example 2-2).

Approximate optimum value of $n = \dfrac{1}{r} = \dfrac{1}{0.1} = 10.$

Number of Haul Units	P_t (Table 4-1)	Expected Production (278 × P_t) (BCY/hr)	Equipment Cost ($/hr)	Cost Performance ($/BCY)
8	0.662	184.0	164	0.8913
9	0.727	202.1	180	0.8906
10	0.785	218.2	196	0.8983
11	0.837	232.7	212	0.9110

Optimum solution $= 9$ trucks @ $0.891/BCY.

Expected production $= 202$ BCY/hr.

Field Observations

As mentioned earlier, field observations indicate that actual loader service rates and haul unit arrival rates do not conform precisely to the Poisson distribution. However, actual production has been found to be close to the predicted values. In one study, predicted production was found to be approximately 3% less than actual production. Thus, improved accuracy may be obtained by multiplying Equation 4-11 by a factor of 1.03, yielding the following equation:

$$\text{expected production} = 1.03 \times \text{excavator production} \times P_t \qquad (4\text{-}12)$$

Correction of the distributions used for the loading rate and arrival rate can best be accomplished by using the simulation methods to be discussed in Chapter 14.

4-4 HAUL UNIT OPERATION

Sizing Haul Units

Previous sections have covered two of the principal methods used for determining the number of haul units of a specific size required to support a specific loader. However, there are other important factors involved in selecting an optimum excavator/haul unit combination. These factors include proper sizing of the haul units, provision for standby units, and management of the haul fleet.

A difficult factor to measure in estimating excavator performance is the effect of the size of the target which the haul unit presents to the excavator operator. It has been found that the use of too small a haul unit will both increase the excavator cycle time and lead to excessive spillage during loading. Job studies have shown that these factors often result in production losses of 10% to 20%. As a rule of thumb, it is suggested that haul units have a minimum capacity of 4 times the excavator bucket capacity. Draglines require even larger target areas. For dragline operations, haul unit sizes of 5 to 10 times bucket capacity are recommended.

It is also desirable to have haul units that hold an integer number of bucket loads. The use of a partially filled bucket to top off a load is always inefficient and is especially costly for haul units that hold a low number of bucket loads (4 or less). Spillage during loading must also be minimized.

Spotting Haul Units

It has been found that careless spotting of haul units at the excavator is one of the most common causes of inefficiency in excavator operations. The location of the loading position of the haul unit should be carefully planned to minimize excavator cycle time. For example, reducing the swing of a shovel by 30° will increase production approximately 15%. The use of a 180°-swing for loading instead of a 90°-swing will reduce shovel production by almost one-third.

The number of spotting operations required per hour is, of course, a function of the haul unit size for a given excavator. Spotting time for back-in loading may be reduced by using spotting logs or bumpers to help the haul unit operator position his vehicle for loading. For shovel loading, haul units should be spotted as close to the bank as possible within the radius of the dipper as it leaves the bank.

To further reduce the loss of production during spotting, it may be advisable to use two spotting positions—one on each side of the loader. In this case, one haul unit may be spotted while the other unit is being loaded. Continuous in-line spotting is also efficient whenever job layout permits and a minimum swing angle can be obtained. Having a supervisor direct spotting may also increase production and be economically feasible.

Standby Units

Since fixed costs for operation of a haul unit are usually small in comparison with those of the excavator, standby haul units are frequently provided to insure that the productive capacity of the excavator is fully utilized. Standby units are used to replace haul units that break down or are unable to perform in synchronization with other haul units.

It is suggested that standby units be provided on a ratio of 1 : 5 for average multiple-excavator operations. This ratio may be reduced for large haul fleets. For single-excavator operations with a small haul fleet, the ratio should be increased to about 1 : 4. If rented equipment is available on short notice, it may be possible to reduce the number of standby units even further. In any case, the exact determination of the number of standby units to be provided should be based on the construction organization's past experience, manufacturer's data on haul unit breakdown, and an analysis of the operational situation.

Management of the Haul Fleet

To obtain maximum efficiency in excavation and hauling operations, the operation of the excavator and haul fleet must be carefully synchronized. Ideally, the haul units would be separated by a fixed distance, travel at exactly the same speed, and arrive at the loader precisely when required for loading. In such an idealized situation the excavator and haul units would all be kept fully utilized. In actual practice, of course, such perfection cannot be obtained. Nevertheless, the objective in managing such an operation is to approach the ideal situation as closely as possible. Techniques that will assist managers in obtaining maximum hauling efficiency are given below.

1. Whenever possible, stagger the starting and stopping time for haul units in order to avoid the bunching up of units at the beginning or the end of a shift.

2. Load haul units as close to the rated load as possible. Do not overload because excessive maintenance and breakdowns will result. Poor haul road conditions may require units to operate at less than rated load. It will usually prove economical to improve the condition of these haul roads.

3. Keep dump bodies clean and in good condition in order to facilitate dumping.

4. Operate haul units at the highest legal safe speed and maintain the desired interval between units. Sluggish units should be replaced by standby units. If standby units are unavailable, load the sluggish unit lightly so that it can maintain the required speed. Do not allow speeding. Speeding is not only unsafe, but it also results in excessive equipment wear and upsets the uniform spacing of haul units.

5. Provide separated haul and return lanes whenever possible. This decreases the chance of accidents and allows higher safe vehicle speeds.

6. Develop an efficient traffic pattern for operation of haul units in loading and dumping areas. Minimize backing and interference between units.

7. Have haul units assist in spreading of fill by spread dumping (dumping while moving forward).

8. Whenever conditions at dump sites are not uniform, alternate haul units between slow and fast dumping sites in order to maintain more uniform intervals between haul units.

9. Use time studies and job observation to determine the factors that limit production. Methods may then be devised to improve the situation. Use actual job data to refine estimates based on average or theoretical data.

PROBLEMS

1. In planning an excavation operation you have available a 3-cu yd wheel loader and 12-cu yd (heaped) dump trucks. The soil is sand and gravel (bucket efficiency = 100%). Job conditions are good and management is excellent. Loader cycle time is estimated at 0.4 min and truck travel time is estimated at 0.3 hr. Use the rational method of estimating truck requirements and production.
   (a) How many dump trucks should be used?
   (b) What is the estimated production in bank cubic yards per hour?

2. Using the traditional method and the shovel of Example 2-2, determine how many trucks holding 14 BCY would be needed to service the shovel if travel time is 0.25 hr?

3. Using queueing theory for a fleet of 7 trucks having a loading time of 0.1 hr and a travel time of 0.4 hr, find the probability of the following:
   (a) A haul unit being available at the loader at any instant.
   (b) No haul unit being available at the loader.

4. Compute the value of P_t for 4 trucks, a loading time of 0.06 hr, and a travel time of 0.6 hr. Use Equation 4-10. Compare the result with the value given in Table 4-1.

5. Using queueing theory and the data from Example 4-4, find the cost performance and expected production of the truck/shovel combinations listed. Use production Equation 4-12.

   (a) What is the optimum number and size of truck to use?
   (a) How do the results compare with the results of the rational method used in Example 4-4?

REFERENCES

1. *Caterpillar Performance Handbook*. Peoria, Illinois: Caterpillar Tractor Co., 1975.

2. HAVERS, JOHN A., and FRANK W. STUBBS, JR., eds., *Handbook of Heavy Construction*. New York: McGraw-Hill, 1971.

3. O'SHEA, J. B., G. N. SLUTKIN, and L. R. SHAFFER, *An Application of the Theory of Queues to the Forecasting of Shovel-truck Fleet Production*. Urbana, Illinois: University of Illinois, 1964.

4. *Technical Bulletin No. 3: Proper Sizing of Excavation and Hauling Equipment*. Milwaukee: PCSA Bureau of Construction Industry Manufacturers Association, 1966.

5

TRACTORS AND DOZERS

5-1 INTRODUCTION

Tractors

The crawler tractor was one of the earliest pieces of selfpowered earthmoving equipment. Augmented by the rubber-tired or wheel-type tractor, it still is one of the most basic and versatile items of equipment in the construction industry. The tractor equipped with a front-mounted blade is known as a *dozer* or *bulldozer*. Originally, the name applied to a particular dozer was derived from the type of blade used on the tractor. Thus, a tractor with a bull blade was called a *bulldozer*, a tractor with an angle blade was called an *angledozer*, etc. Today, however, all tractors with blades are commonly called either dozers or bulldozers. The types of blades used with tractors, the application of dozers to earthmoving work, and methods of production estimation will be discussed later in this chapter.

In addition to their dozer applications, tractors are used for towing scrapers, wagons and compaction equipment, and for ripping and scarifying. They may also be equipped with front-end loader attachments. These applications are discussed in other chapters. Tractors are usually powered by heavy-duty diesel engines, many of which are turbocharged. Both manual and automatic transmissions are available. Control systems may be either hydraulic or mechanical. Current trends include the use of hydraulic controls and automatic transmissions.

77

Crawler Tractors

The crawler-type tractor has excellent all-terrain versatility because of its low ground bearing pressure (about 6 to 9 lb/sq in.) and excellent traction. Special low ground pressure models are available which have ground contact pressures on the order of 3 to 4 psi. Crawler tractors can operate on side slopes up to 100% (45°). Because of their relatively slow speed, however, they must be transported on equipment trailers for long moves.

The track used on crawler tractors consists of linked shoes of heat-treated steel designed to resist wear. The track runs on rollers mounted on the track roller frame. An idler wheel is mounted on the front of the frame and is equipped with a recoil device with adjustable tension in order to maintain proper track tension and to absorb heavy shocks. The driving sprocket wheel is mounted at the back of the track roller frame. Since track rollers are lubricated and protected by seals to keep out water and abrasives, crawler tractors may operate in water as deep as the height of the track. If it is properly waterproofed, a tractor may be operated for short periods of time in even deeper water.

Dual tractors that are mechanically coupled one behind the other and controlled by a single operator have been developed for use whenever great power is

Figure 5-1   Crawler tractor dozer. (Courtesy Caterpillar Tractor Co.)

required, such as for ripping and for push-loading scrapers. Twin and side-by-side tractors have also been developed that provide twice the power of single units and are equipped with a single blade 24 ft or so in width. A typical crawler tractor dozer is shown in Figure 5-1.

Wheel Tractors

Rubber-tired or wheel tractors were developed to yield higher speed in towing scrapers, wagons, and similar equipment. They are available in both 2-wheel and 4-wheel models. Two-wheel models must be operated with specially designed matching components (scrapers, etc.) in order to maintain their balance. Four-wheel tractors are available in either 2-wheel or 4-wheel drive models and may be used to tow any type of equipment. Hydraulic control systems and automatic transmissions are typically used on wheel tractors.

Wheel tractors may be equipped with dozer blades or any of the other attachments previously mentioned for crawler tractors. However, the wheel tractor's ability to perform dozing is limited by its traction and comparatively high ground pressure (typically 25 to 35 psi). Traction and tractor performance will be discussed in a succeeding section. In addition to possessing high travel speeds, wheel tractors may operate on paved roads without damaging the surface. Another advantage of the wheel tractor is its ability (because of its high ground pressure) to assist in compacting soils. An articulated rubber-tired dozer is shown in Figure 5-2.

Figure 5-2 Articulated rubber tired dozer. (Courtesy Clark Equipment Co.)

5-2 BLADES AND ATTACHMENTS

Dozer Blades

Although there are a number of different types of dozer blades available, the four most common are the straight blade, the angle blade, the universal blade, and the cushion blade. All of these except the cushion blade may be tilted laterally to permit the concentration of tractor power on one end of the blade. Tilting is especially useful in cutting ditches and in breaking up soils that have a hard crust. Both the straight and universal blades allow the top of the blade to be moved forward or backward (pitched). Pitching the blade increases or decreases blade penetration by varying the angle of attack of the blade against the ground. Only the angle blade may be turned so that it is not perpendicular to the direction of travel of the tractor. Pitching, tilting, and angling are illustrated in Figure 5-3.

Tilting   Pitching   Angling

Figure 5-3   Dozer blade adjustments.

Two terms used to indicate the potential performance of a tractor and blade combination are *horsepower per foot of cutting edge* and *horsepower per loose cubic yard*. The horsepower per foot of cutting edge gives an indication of the blade's ability to penetrate a soil and obtain a full blade load of material. The horsepower per loose cubic yard is a measure of the blade's ability to move material once the blade is loaded.

The straight blade (Figure 5-4) is generally considered the most versatile dozing blade. It is best used for general dozing when material is being moved a short or medium distance. It has a higher horsepower per foot of cutting edge and a higher horsepower per loose cubic yard than does the universal blade. Thus, it has a higher penetrating ability and can push heavier material than can the universal blade. It can be equipped with a push plate for push-loading of scrapers. Bull blades are similar to straight blades, but they cannot be tilted. Bull blades are now seldom used.

Angle blades may be positioned straight or angled approximately 25° to either side of the straight position. The blade may be tilted but not pitched. Angle blades are most effectively used for making sidehill cuts, backfilling, and ditching. They have also been used successfully for rough grading and for windrowing (moving material laterally in a longitudinal pile).

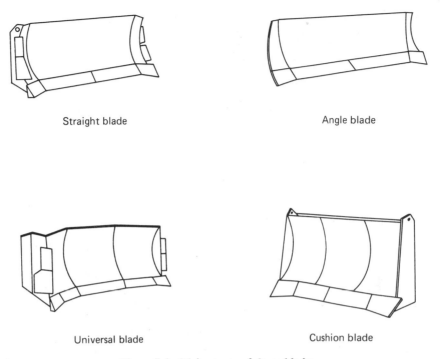

Straight blade

Angle blade

Universal blade

Cushion blade

Figure 5-4 Major types of dozer blades.

The universal blade with its wings on each end of the blade can move large volume loads over long distances. Since it has low horsepower per foot of cutting edge and per loose cubic yard, it should not be used to obtain high penetration or to move heavy materials.

The cushion blade is designed primarily for push-loading of scrapers. It is reinforced and cushioned to absorb the shock of contacting the push block of scrapers. It may also be used for general-purpose dozing and for cleanup work.

Some of the other types of dozer blades available include the ripdozer, the light material U-blade, and special-purpose clearing blades. The ripdozer is similar to a straight blade but has adjustable shanks at each end which can extend up to 12 in. below the cutting edge. Thus, it is able to penetrate and load tougher soils than other blades. It is also useful for removing stumps, breaking up frozen soils, and assisting rippers.

Light material U-blades are similar to universal blades, but they are of lighter construction and carry a larger volume of material. The light material U-blade is most suitable for moving high volumes of light, noncohesive materials.

Special-purpose clearing blades are designed for clearing brush and trees rather than for earthmoving. The two general special-purpose clearing blades are the V-type blade and the angling blade. Both are equipped with a very sharp cutting edge for severing brush and trees at or just below the ground line. The angling blade is permanently mounted at an angle to the dozer with a knife type extension or stinger on the front end of the blade. The stinger is used to split larger trees before the blade cuts them off and pushes them over.

Other Attachments

In addition to the blades mentioned above, other common attachments for tractors and dozers include rakes, plows, scarifiers, and rippers. Rakes are used in land clearing to rapidly gather up brush and small fallen trees. A heavier rake is available for combing large rock out of land being cleared for agricultural use. Plows, scarifiers, and rippers are used to break up tough soils for excavation. A discussion of ripper operation is contained in Chapter 9.

5-3 DETERMINING TRACTOR SPEED

General

The maximum speed attainable under specific load and operating conditions by a tractor, scraper, or truck is determined by the power available at the wheel or track, the limitations of traction, and the retarding forces of grade and rolling resistance. Methods of evaluating these factors and determining the vehicle's maximum speed from the equipment manufacturer's performance curves will be presented in this section.

Resistance to Movement

The total resistance to movement of a vehicle over a surface is the sum of its rolling resistance and the grade resistance.

$$\text{total resistance} = \text{rolling resistance} + \text{grade resistance} \qquad (5\text{-}1)$$

Total resistance is normally expressed in pounds. However, resistance is also expressed in pounds per ton or in an equivalent percentage of grade, as will be seen shortly.

Rolling Resistance

Rolling resistance is the resistance to movement of a vehicle over the travel surface due to internal friction, tire flexing, and penetration of the vehicle into the travel surface. It has been found that the rolling resistance of a rubber-tired vehicle on a hard, smooth, level surface amounts to approximately 40 lb/ton of vehicle weight with bias or belted/bias tires and as little as 30 lb/ton with radial tires. It has also been observed that rolling resistance increases approximately 30 lb/ton for each inch of tire penetration into the travel surface. This leads to the following equations:

$$\text{rolling resistance (lb/ton)} = 40 + (30 \times \text{inches penetration}) \qquad (5\text{-}2)$$

$$\text{rolling resistance (lb)} = \text{rolling resistance (lb/ton)} \times \text{weight (tons)} \qquad (5\text{-}3)$$

Rolling resistance in pounds per ton for some typical surface conditions are given in Table 5-1.

TABLE 5-1 *Typical values of rolling resistance*

Type of Road Surface	Rolling Resistance (lb/ton)
Concrete or asphalt	40 (30)*
Firm, smooth, flexing slightly under load	65 (50–55)*
Unplowed dirt or muddy surface on firm base	80
Snow, packed	50
Snow, loose	90
Rutted dirt roadway, 1 in. to 2 in. penetration	100
Soft, rutted dirt, 3 in. to 4 in. penetration	150
Loose sand or gravel	200
Soft, muddy, rutted	300–400

*Radial tires

Since crawler tractors are in effect traveling over their own road (interior surface of their tracks), they are often considered to have no rolling resistance. Actually, there is some variation in rolling resistance over different types of surfaces, but this is usually neglected in calculating resistance. However, when crawler tractors are towing wheeled equipment, the rolling resistance of the wheeled equipment must be considered when computing total resistance of the combinations. The rolling resistance of the surface on which the rated drawbar pull of crawler tractors is determined is actually 110 lb/ton.

Grade Resistance

Grade resistance is the component of vehicle weight that acts parallel to an inclined travel surface. This is, of course, a positive resistance for vehicles traveling uphill and a negative resistance (or propelling force) for vehicles traveling downhill. The exact value of grade resistance is obtained by multiplying the weight of the vehicle by the sine of the angle that the roadway makes with the horizontal. However, for small angles, the sine of the angle is approximately equal to the tangent of the angle. For a 15°-slope, this approximation results in an error of less than 3%. Grade is usually expressed in percent (that is, a 1%-grade has a rise or fall of 1 ft in 100 ft), which is the same as the tangent of the slope angle. Thus, for grades usually encountered in construction operations, it is sufficiently accurate to use the following equations:

$$\text{grade resistance (lb)} = \text{weight (lb)} \times \text{grade} \qquad (5\text{-}4)$$

$$\text{grade resistance (lb/ton)} = 20 \times \text{grade (\%)} \qquad (5\text{-}5)$$

Effective Grade

The total resistance of a vehicle may also be expressed as the percent grade that would have a grade resistance equal to the total resistance computed. This is referred to as *effective grade*, *equivalent grade*, or *percent total resistance*. Since a 1%-grade is equivalent to a resistance of 20 lb/ton,

$$\text{effective grade (\%)} = \frac{\text{total resistance (lb/ton)}}{20} \qquad (5\text{-}6)$$

or

$$\text{effective grade (\%)} = \text{grade (\%)} + \frac{\text{rolling resistance (lb/ton)}}{20} \qquad (5\text{-}7)$$

The calculation of total resistance as an equivalent effective grade is illustrated in the following examples.

Example 5-1

PROBLEM: A wheel tractor is being operated on a soft roadway with a tire penetration of 5 in. The tractor weighs 20 tons. What is the total resistance and effective grade when
(a)   The tractor is ascending a slope of 4%?
(b)   The tractor is descending a slope of 6%?

Solution:

Rolling resistance factor $= 40 + (30 \times 5) = 190$ lb/ton (Equation 5-2).

Rolling resistance $= 190 \times 20 = 3,800$ lb (Equation 5-3).

(a)   Grade resistance $= 0.04 \times 20 \times 2,000 = 1,600$ lb (Equation 5-4).
Total resistance $= 3,800 + 1,600 = 5,400$ lb (Equation 5-1).
Effective grade $= \dfrac{5,400}{20 \times 20} = 13.5\%$ (Equation 5-6) or effective grade $=$
$4 + \dfrac{190}{20} = 4 + 9.5 = 13.5\%$ (Equation 5-7).

(b)   Grade resistance $= -0.06 \times 20 \times 2,000 = -2,400$ lb (Equation 5-4).
Total resistance $= 3,800 - 2,400 = 1,400$ lb (Equation 5-1).
Effective grade $= -6.0 + \dfrac{190}{20} = -6.0 + 9.5 = 3.5\%$ (Equation 5-7).

Example 5-2

PROBLEM: A crawler tractor weighing 60,000 lb is towing a rubber-tired scraper weighing 48,000 lb up a grade of 5%. The rolling resistance of the haul road is 100 lb/ton. What is the total resistance of the combination?

Solution:

Rolling resistance (neglect crawler) $= \dfrac{48,000}{2,000} \times 100 = 2,400$ lb (Equation 5-3).

Grade resistance $= (60,000 + 48,000) \times 0.05 = 5,400$ lb (Equation 5-4).

Total resistance $= 2,400 + 5,400 = 7,800$ (Equation 5-1).

Power Available

Before using equipment performance curves to determine maximum vehicle speed, it is necessary to consider two factors that may limit the usable power of the vehicle. These factors are altitude and traction.

Diesel engines used in construction equipment produce less power as the altitude of operation is increased because of the decreased density of air at increased altitude. Turbocharged engines are more efficient than naturally aspirated engines in this respect and may maintain their rated power up to altitudes of 10,000 ft or so. Manufacturers publish data, usually in the form of derating tables, on the performance characteristics of their machines at various altitudes.

For naturally aspirated engines, there is some difference between the altitude performance of two-cycle and four-cycle diesel engines. Various engine manufacturers also rate their engines differently for altitude performance. However, for estimating purposes when derating tables are unavailable, it is sufficiently accurate to reduce the rated performance of the equipment by 3% for each 1,000 ft of altitude above 3,000 ft.

$$\text{derating factor } (\%) = 3 \times \frac{\text{altitude} - 3,000}{1,000} \qquad (5\text{-}8)$$

The percentage of rated power available is, of course, one hundred minus the derating factor. Thus, the maximum available pull is calculated by multiplying the rated pull by the percent of power available, as illustrated in Example 5-3. The use of derating factors in estimating vehicle speed and travel time will be illustrated in Example 5-5.

The power produced by a tractor is expressed as *drawbar pull* for crawler tractors and *rimpull* for wheel tractors, scrapers, and trucks. Drawbar pull is the pulling force available at the tractor hitch when a crawler tractor is tested under standard conditions. Rimpull is the pull available at the rim of the driving wheels when wheel tractors are operated under standard conditions. Usable drawbar pull and rimpull are limited by the traction which can be developed by the track or tires. The maximum usable pull of a tractor is determined by the following equation:

$$\text{maximum usable pull} = \text{coefficient of traction} \times \text{weight on drivers} \quad (5\text{-}9)$$

Approximate values of coefficient of traction for typical surfaces are given in Table 5-2.

TABLE 5-2 *Typical values of coefficient of traction*

Type Surface	Rubber Tires	Tracks
Concrete, dry	0.90	0.45
Concrete, wet	0.80	0.45
Earth or clay loam, dry	0.60	0.90
Earth or clay loam, wet	0.45	0.70
Gravel, loose	0.35	0.50
Quarry pit	0.65	0.55
Sand, dry, loose	0.25	0.30
Sand, wet	0.40	0.50
Snow, packed	0.20	0.25
Ice	0.10	0.15

To determine the weight on the drivers, use total tractor weight for crawler tractors and 4-wheel drive wheel tractors. For tractor-scrapers, use manufacturer's specifications if they are available. When specifications are not available, use 40% of gross vehicle weight (tractor plus scraper) for 4-wheel tractors and 60% of gross vehicle weight for 2-wheel tractors.

The application of altitude and traction considerations to tractor performance is illustrated in the following example.

Example 5-3

PROBLEM: A 4-wheel drive wheel tractor weighs 41,000 lb and produces a maximum rimpull of 40,000 lb. It is located at an altitude of 8,000 ft on wet earth. Operating conditions require a pull of 20,000 lb to move the tractor and its load. Can the tractor perform under these conditions?

Solution:

Derating factor $= 3 \times \left(\dfrac{8,000 - 3,000}{1,000}\right) = 15\%$ (Equation 5-8).

Percent rated power available $= 100 - 15 = 85\%$.

Available power $= 40,000 \times 0.85 = 34,000$ lb.

Coefficient of traction $= 0.45$ (Table 5-2).

Maximum usable pull $= 0.45 \times 41,000 = 18,450$ lb (Equation 5-9).

Usable pull $(18,450) <$ required pull $(20,000)$.

$\therefore$ Tractor cannot perform under these conditions.

Using Performance and Retarder Curves

After total resistance, derating factors, and traction limitations have been determined, equipment performance and retarder curves may be used to determine the maximum speed at which the vehicle can operate under these conditions.

Performance and retarder curves are similar in most respects. However, a performance curve reflects engine power and transmission charateristics so that it indicates the maximum speed a vehicle can maintain while supplying a certain drawbar pull or rimpull (total resistance positive). Retarder curves indicate the safe speed range in which the vehicle may operate while descending steep grades without using its brakes (total resistance is negative). Retarder curves derive their name from the hydraulic retarder that is used on the vehicle to absorb horsepower through the transmission cooling system. The total braking horsepower available as shown on the retarder curve is actually a combination of retarder braking power and engine friction with a closed throttle. The hydraulic retarder may be used to slow a vehicle traveling on level ground, but it will not provide a positive stop.

Performance Curves

A relatively simple performance curve of the type often used for tractors is illustrated in Figure 5-5. Rimpull or drawbar pull is shown on the vertical scale

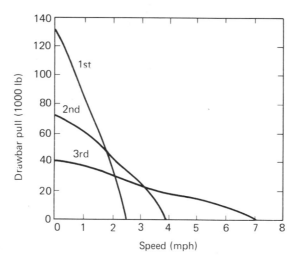

Figure 5-5 Typical crawler tractor performance curve.

and vehicle speed is shown on the horizontal scale. To use this curve, enter on the vertical scale at the required pull (total resistance), move horizontally to the right until the performance curve for a particular gear is intersected, and then drop vertically to read the maximum speed on the bottom scale. When the horizontal line crosses performance curves for two or more gears, the intersection point farthest to the right would normally be used since this represents the maximum possible speed of the vehicle under the conditions.

Example 5-4

PROBLEM: Using the performance curve of Figure 5-5, determine the maximum operating speed of the tractor when the required pull (total resistance) is 60,000 lb.

Solution:

Entering Figure 5-5 at a drawbar pull of 60,000 lb, move horizontally to the right until you intersect the curves for first and second gear. Read the corresponding speeds (first gear = 1.5 mph: second gear = 1.0 mph). Thus, the maximum speed is found to be about 1.5 mph using first gear.

A more complex performance curve used by many manufacturers for trucks and tractor-scrapers provides for graphical calculation of required vehicle pull in addition to indicating maximum speed. This performance curve is illustrated in Figure 5-6. To utilize this chart, start with the gross vehicle weight (vehicle plus load) on the top scale. Drop vertically until you intersect the diagonal line corresponding to the total resistance (in percent of vehicle weight-same as effective grade). Move horizontally across to the appropriate scale and read the total pull required under the operating conditions. When altitude is a consideration, the total

631C RIMPULL CURVE

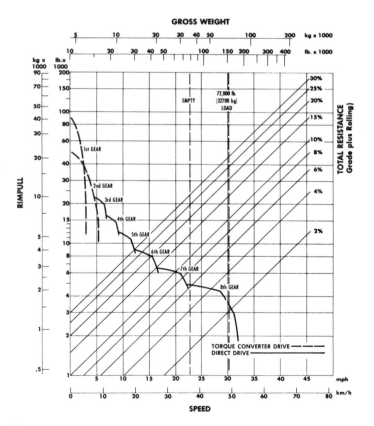

Figure 5-6 Wheel scraper performance curve. (Courtesy Caterpillar Tractor Co.)

resistance must be adjusted for altitude prior to entering the performance chart. This is accomplished by dividing the total resistance (in pounds) by the quantity "1 — derating factor (expressed as a decimal)" to obtain the required rimpull or drawbar pull. The maximum vehicle speed and appropriate gear are then found in the same manner as described for the use of Figure 5-5.

Example 5-5

PROBLEM: Using the performance curve of Figure 5-6, determine the maximum speed of the vehicle at its rated gross vehicle weight, an effective grade of 10%, and an altitude derating factor of 25%.

Solution:

Start on the top scale and drop down the rated gross vehicle weight line until it intersects the 10%-total resistance line. Projecting this horizontally to the left scale, you will find that the required rimpull is approximately 15,000 lb. Dividing by (1 — 0.25 = 0.75) yields an equivalent required rimpull of 20,000 lb at rated altitude. Move horizontally to the right from 20,000 lb required pull until you intersect the performance curve. Drop vertically and you will find a maximum speed of about 6 mph in third gear.

A retarder curve is illustrated in Figure 5-7. This chart is used in a similar manner to that of Figure 5-6. However, remember that the total resistance in this

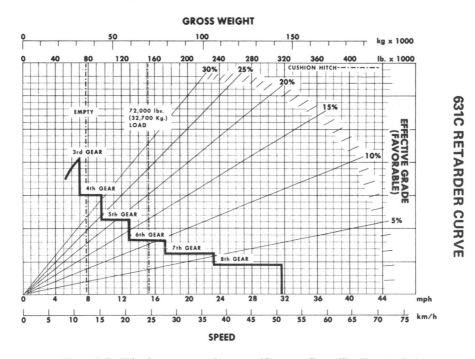

Figure 5-7 Wheel scraper retarder curve. (Courtesy Caterpillar Tractor Co.)

case represents a negative resistance or negative effective grade. The horizontal line representing required rimpull (negative) will normally intersect two or more gear curves. The vehicle may operate at any speed between the maximum and minimum values by changing the transmission range, varying the amount of oil in the retarder, or by accelerating the engine. As with the performance curve, the highest speed which is safe for the operating conditions would normally be used.

The speeds obtained from performance and retarder curves do not consider acceleration, deceleration, or gear changes. Methods for calculating tractor and truck travel times to include these factors will be covered in Chapter 6.

5-4 ESTIMATING DOZER PRODUCTION

General

The production of dozers, as of most other earthmoving equipment, may be estimated by multiplying the quantity of material moved per cycle by the expected number of cycles per hour (including a job efficiency factor). Dozer production may also be estimated by using production curves provided by some dozer manufacturers. In either case, dozer production is usually estimated in loose cubic yards per hour and converted to bank cubic yards when required.

Blade Capacity and Cycle Time

In order to calculate dozer production by this method, it is necessary to estimate the average blade load that will be obtained on the job. Basically, there are four methods of accomplishing this:

1. Use the blade manufacturer's rating of blade capacity.

2. Base estimate on previous experience with similar materials, equipment, and job conditions.

3. Weigh blade loads obtained in field tests.

4. Measure blade loads obtained in field tests.

The first three methods are self-explanatory. The following procedure is suggested for measuring the blade load:

1. Obtain a full blade load, carry onto a level surface, and lift the blade while pulling slightly forward so that an evenly shaped pile is formed.

2. Obtain the width of the pile (W) in feet perpendicular to the blade by averaging two measurements taken in line with the inside edge of each track or wheel.

3. Obtain the height of the pile (H) in feet by averaging two height measurements taken in a similar manner.

4. Obtain the length (L) of the pile parallel to the blade in feet.

5. Calculate blade volume by using the following equation:

$$\text{blade load (LCY)} = 0.0139 \times H \times W \times L \qquad (5\text{-}10)$$

Once the average blade load has been determined, the average cycle time must be estimated. Total cycle time is the fixed cycle time plus variable cycle time. Fixed cycle time accounts for the time required to spot and start loading as well as change gears. The values given in Table 5-3 are suggested for dozer fixed cycle time.

TABLE 5-3 *Dozer fixed cycle time*

Operating Conditions	Time (min)
Power shift transmission, average	0.05
Direct drive transmission, average	0.10
Hard digging	0.15

Variable time is the time required for dozing and returning. Since a dozer normally moves a load only a short distance, return is usually in reverse. It is suggested that the speeds given in Table 5-4 be used in calculating travel times.

TABLE 5-4 *Dozer operating speeds*

Operating Conditions		Direct Drive	Power Shift
Dozing:			
Hard materials, haul 100 ft or less		first gear	1.5 mph
Hard materials, haul over 100 ft		second gear	2.0 mph
Loose materials, haul 100 ft or less		second gear	2.0 mph
Loose materials, haul over 100 ft		third gear	2.5 mph
Return:			
100 ft or less	Reverse speed in gear used for dozing		Maximum reverse speed in second range
Over 100 ft	Highest reverse speed		Maximum reverse speed in third range

Dozing and return times are calculated by dividing travel distance by dozing and return speeds, respectively.

Example 5-6

PROBLEM: A power shift crawler tractor has a rated blade capacity of 5 LCY. It is being used to push loose common earth an average distance of 80 ft. Maximum reverse speed in second range is 4.6 mph. What is the estimated production of this dozer per 50-min hour?

Solution:

Conversion factor: 1 mph = 88 fpm.

Fixed time = 0.05 min (Table 5-3).

Haul time = $\dfrac{80}{88 \times 2.0}$ = 0.45 min (Table 5-4).

Return time = $\dfrac{80}{88 \times 4.6}$ = 0.20 min (Table 5-4 and speed data).

Total cycle time = 0.70 min.

Production = $\dfrac{5 \times 50}{0.70}$ = 357 LCY/hr.

Using Production Curves

Some manufacturers provide production curves such as those shown in Figure 5-8 for their dozers. To utilize these curves for production estimating, the maximum uncorrected production determined from the appropriate curve is multiplied by the appropriate correction factors and the job efficiency factor.

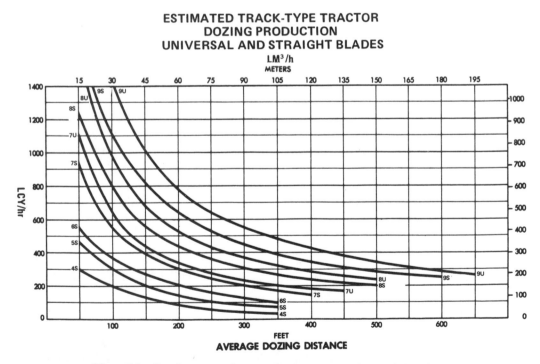

Figure 5-8 Crawler tractor dozer production curve. This chart is based on numerous field studies made under varying conditions and jobs. (Courtesy Caterpillar Tractor Co.)

Correction Factors

Job Condition Corrections:	Track-Type Tractor	Wheel-Type Tractor
Operator: Excellent	1.00	1.00
Average	0.75	0.60
Poor	0–0.60	0–0.50
Material:		
Weight:		
$\dfrac{3000\ \text{lb/BCY}}{\text{actual lb/BCY}}$ or $\dfrac{2300\ \text{lb/LCY}}{\text{actual lb/LCY}}$		
Type:		
Loose stockpile	1.20	1.20
Hard to cut; frozen		
with tilt cylinder	0.80	0.75
without tilt cylinder	0.70	—
cable controlled blade	0.60	—
Hard to drift; "dead" (dry,		
non-cohesive material or		
very sticky material)	0.80	0.80
Rock, ripped or blasted	0.60–0.80	—
Slot Dozing	1.20	1.20
Side by Side Dozing...........	1.15–1.25	1.15–1.25
Visibility: Dust, rain, snow,		
fog or darkness	0.80	0.70
Job Efficiency: 50 min/hr	0.84	0.84
40 min/hr	0.67	0.67
Direct Drive Transmission		
(0.1 min. fixed time)	0.80	—
*Bulldozer: Angling (A) blade ..	0.50–0.75	—
Cushioned (C)		
blade............	0.50–0.75	0.50–0.75
Rip (R) Blade	1.00–1.50	—
D5 narrow gauge ..	0.90	—
Light material		
U-blade (coal)....	1.20	1.20
Blade bowl		
(stockpiles)	1.30	1.30

Grades:

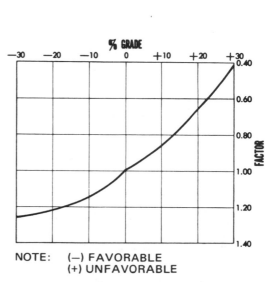

NOTE: (−) FAVORABLE
       (+) UNFAVORABLE

*Note: Angling blades and cushion blades are not considered production dozing tools. Depending on job conditions, the A-blade and C-blade will average 50–75 % of straight blade production.

Figure 5-8 (*cont.*)

Example 5-7

PROBLEM: Using the production curves of Figure 5-8, determine the expected output of a D-7 crawler-type dozer with straight blade (curve 7S) when dozing sand down a 5%-grade for an average distance of 100 ft. Operator is average. Job efficiency is estimated at 50 min/hour. Estimated material weight is 2,400 lb/LCY.

Solution:

Maximum uncorrected production = 550 LCY/hr (Figure 5-8).

Correction factors (Figure 5-8).

Hard to drift material = 0.80.

$$\text{Material weight} = \frac{2{,}300}{2{,}400} = 0.96.$$

Average operator $= 0.75$.

Grade $(-5\%) = 1.07$.

Job efficiency $= 0.83$.

Expected production $= 550 \times 0.80 \times 0.96 \times 0.75 \times 1.07 \times 0.83$.

Expected production $= 281$ LCY/hr.

5-5 JOB MANAGEMENT

Use of Dozers

The earthmoving application of dozers principally involves excavating and pushing soil over relatively short distances. Operating conditions may include rough terrain, steep slopes, and poor traction. This short-distance zone of operation is often referred to as the *power zone* of earthmoving. While the maximum efficient dozing distance is often considered to be only 300 ft or so, this distance is actually determined by dozer and job characteristics. The efficient dozing range may be extended by using the techniques described in the following paragraph and may extend to 1,000 ft or more when employing large, mechanically coupled, side-by-side dozers.

Techniques to Increase Production

In addition to choosing the most appropriate tractor and blade combination for the expected job conditions, other techniques may be used to increase dozer production. These include downhill dozing, slot dozing, and blade-to-blade (or side-by-side) dozing.

Downhill dozing may greatly increase dozer production, as can be seen from the grade correction graph of Figure 5-3. It is not necessary for the dozer to actually travel downhill on each pass to take advantage of the production increase that results from downhill dozing. Under some conditions it may be more efficient to pile up several blade loads at the brink of the hill and then push them to the bottom of the hill in one pass.

Slot dozing utilizes spillage from the initial dozer passes to form ridges on each side of the dozer's cut area. Thus, a slot or trench is created which greatly increases the load that the blade can carry to the dump area. Slot dozing may increase dozer production up to 50% under favorable grade and soil conditions. The technique may be applied to large cut areas by leaving narrow uncut sections between slots. The uncut sections are later dozed out near the end of the excavation process.

Blade-to-blade dozing utilizes two or more dozers operated in parallel with their blades almost touching. Again, this enables the blade load to be increased considerably. However, it is usually not efficient for dozing distances of less than 50 ft since the increase in blade load is offset by the extra maneuvering time required. The use of side-by-side (SxS) dozers in which two dozers are mechanically coupled, have a single blade, and use a single operator is more efficient than using blade-to-blade dozing.

Traction

Traction conditions affect the operating efficiency of both crawler and wheel dozers. However, the effect on production is more pronounced with wheel dozers. Dozer production curves such as shown in Figure 5-8 assume some minimum coefficient of traction and must be adjusted when the coefficient of traction is less than assumed. A rule of thumb to use is that wheel dozer production decreases about 4% for each 0.01 decrease in the coefficient of traction below 0.40.

Job Efficiency

Job efficiency for dozer production may be estimated by using Table 2-4 (which considers job conditions and management factors) or by estimating the number of actual working minutes per hour. When historical data from previous jobs under similar conditions are not available, the job efficiency factors of Table 5-5 are suggested.

TABLE 5-5 *Average dozer efficiency factors*

Operating Conditions	Type Tractor	Efficiency Factor
Day	Crawler	0.83
Day	Wheel	0.75
Night	Crawler	0.75
Night	Wheel	0.67

PROBLEMS

1. What type of dozer blade would you select for the following applications?
   (a) Excavating a V-type ditch.
   (b) Breaking up frozen soil.
   (c) Making a sidehill cut for a road.
   (d) Push-loading a scraper.

2. What is an angledozer?

3. A wheel tractor weighing 72,000 lb is pulling a wagon weighing 100,000 lb up a 5%-grade. What is the total resistance of this combination if the haul route has a rolling resistance of 80 lb/ton?

4. An off-highway truck weighs 60,000 lb empty and can carry a payload of 100,000 lb. The haul route requires the truck to travel down a 3%-grade and return empty on the same route. The haul road has a rolling resistance of 100 lb/ton. What are the total resistance and effective grade for each portion of the haul route?

5. A naturally aspirated tractor is operating at an altitude of 12,000 ft. What derating factor should be used if the manufacturer's data are unavailable?

6. The wheel tractor of Problem 3 is equipped with 4-wheel drive. What is the maximum grade this tractor-wagon combination could ascend if the haul road surface consisted of loose gravel?

7. The tractor-scraper whose performance curve is shown in Figure 5-6 is operating at an altitude at which the derating factor is 6%. What is the maximum speed of this vehicle at rated load when ascending a grade of 10%? The rolling resistance is 100 lb/ton.

8. The coefficient of traction of the surface over which a wheel dozer is operating is 0.30. What percent of normal production would this dozer be expected to achieve?

9. In a field trial to determine blade load you measure dimensions of the pile to be: length = 17 ft, width = 8 ft, and height = 6 ft. The crawler dozer used has power shift. Average dozing distance is estimated at 200 ft. You classify the material as loose. Estimate the production of this dozer during night operations. Dozer reverse speeds are:

Gear Range	Speed (mph)
1	0–3.0
2	0–5.5
3	0–8.0

10. Use the 9U production curve of Figure 5-8 to estimate the production of a dozer under the following conditions:
   (a) Operator = average.
   (b) Material = sticky clay weighing 2,700 lb/LCY.
   (c) Grade = +5%.
   (d) Job efficiency = 0.83.
   (e) Average dozing distance = 150 ft.

REFERENCES

1. *Basic Estimating* (3rd ed.). Melrose Park, Illinois: Construction Equipment Division, International Harvester Company, n.d.

2. *Caterpillar Performance Handbook*. Peoria, Illinois: Caterpillar Tractor Co., 1975.

3. *Engineering Bulletin RB-224A: Road Building Equipment*. Chicago: American Oil Company, 1962.

4. *Production and Cost Estimating of Material Movement with Equipment*. Hudson, Ohio: TEREX Division, General Motors Corporation, 1970.

6

SCRAPERS

6-1 INTRODUCTION

Characteristics and Nomenclature

Scrapers are designed for loading, hauling, and dumping soil over medium to long distances in the earthmoving process. Although they are often referred to as self-loading because they do not require the services of an excavator or loader, only under certain conditions can they load without the assistance of a pusher tractor. The use of pushers and other auxiliary equipment will be discussed later in this chapter.

The three basic operating components of a scraper are the bowl, apron, and tailgate (or ejector). The bowl is the load-carrying portion of the body. Practically all scrapers made today are open-bowl; that is, there is no obstruction over the top of the bowl. The bottom front of the bowl is equipped with a replaceable cutting edge. The bowl can be lowered to obtain the desired depth of cut or raised to clear the ground while hauling. The apron forms the front wall of the bowl and can be raised or lowered as desired. In the elevating scraper the elevator replaces the apron of the conventional scraper. The tailgate or ejector usually forms the rear wall of the bowl and is moved forward to eject the load. However, the ejector sometimes consists of the rear wall and floor of the bowl which is moved upward and forward to dump the load.

Types of Scrapers

Scrapers may be towed by standard crawler or wheel tractors or they may form a part of wheel tractor-scrapers combinations. Most scrapers have only a single axle so that a portion of the scraper and load weight is carried by the tractor. When used with a wheel tractor this serves to increase the weight on the driving wheels so that traction is increased. However, four-wheel (two-axle) scrapers are used with crawler tractors because the hitch of these tractors is not designed to carry a vertical load.

Scrapers pulled by wheel-type tractors may be classified as

1. Single-engine, overhung (or two-axle)
2. Three-axle
3. All-wheel drive
4. Tandem-powered
5. Elevating
6. Push-pull
7. Multi-bowl, multi-engine

A twin-powered, all-wheel drive scraper is shown in Figure 6-1; a twin-engine elevating scraper appears in Figure 6-2.

Figure 6-1 Twin-powered all-wheel drive scraper. (Courtesy TEREX Division, General Motors Corp.)

Figure 6-2 Twin-engine elevating scraper. (Courtesy WABCO Construction & Mining Equipment Group.)

Single-engine, overhung or two-axle scrapers use a tractor that has only one axle. This tractor has certain performance advantages which will be covered later, but it cannot, of course, operate without a matching scraper or other trailing component. The three-axle tractor-scraper is pulled by a four-wheel tractor similar to those discussed in the previous chapter. All-wheel drive scrapers have driving wheels on the scraper as well as on the tractor. Tandem-powered scrapers are all-wheel drive scrapers that have separate engines for powering the scraper wheels. Elevating scrapers use a powered ladder-type elevator to assist in cutting and lifting the soil into the scraper bowl. In most cases, elevating scrapers load without assistance. Push-pull scrapers are equipped with hooks on the rear and bails on the front so that they may couple together and assist each other in loading. Details of their operation will be covered later in the chapter. Multi-bowl, multi-engine scrapers, as their name implies, are specially designed units using a number of powered scraper bowls coupled together to form a sort of scraper train. They are capable of rapidly moving large quantities of earth efficiently under favorable conditions, but details of their operation will not be covered here because of their somewhat specialized nature.

Cutting Edges

Scraper cutting edges usually consist of several sections which bolt to the front edge of the bowl. In the stinger arrangement the center section extends beyond the outer sections. In the level cut arrangement all edge sections extend the same distance. The stinger arrangement is standard on most scrapers because it provides better penetration and more cutting edge wear material than the level cut arrangement. Curved cutting edges are sometimes used for the same reasons. The level cut arrangement is preferred for finish work or when edge breakage is a problem. Some cutting edges are available with integral cutting teeth for high impact applications. Elevating scrapers often use separate teeth attached to the bowl cutting edge. Replaceable router bits are often used to protect the corners of the scraper bowl during operation. Standard and heavy-duty router bits are available.

6-2 SCRAPER PRODUCTION

The Scraper Cycle

As with most earthmoving operations, scraper production is estimated by multiplying the average load per cycle by the number of cycles expected to be completed per unit of time. In order to estimate the number of cycles which will be obtained per hour, the average time required per cycle must be determined. The total cycle time for a scraper is the sum of the fixed cycle time and the variable cycle time.

$$\text{total cycle time} = \text{fixed cycle time} + \text{variable cycle time} \qquad (5\text{-}1)$$

Fixed cycle time and variable cycle time are broken down into the following components:

Fixed cycles:	Load time
	Maneuver and dump time
	Spot and delay time
Variable cycle:	Haul time
	Return time

Fixed Cycle Time

Scraper loading time depends on a number of factors, for example, size and type of scraper, amount of pusher power available, soil type and condition, grade and condition of the loading area, and operator skill. The best estimate of loading time is usually obtained by timing field tests or by using the results of previous operations under similar conditions. However, observation of a large number of jobs indicates that average loading time for push-loaded scrapers range from about 0.6 to 1.0 min. Table 6-1 gives typical loading times for various equipment and job combinations. The optimum loading time for a particular job may be determined by the methods of Section 6.4.

TABLE 6-1 *Scraper loading times* (*min*)

Job Conditions	Single-engine, Overhung (Push-load)	3-Axle (Push-load)	All-wheel Drive		Tandem		Elevating (Self-load)
			Push-load	Self-load	Push-load	Self-load	
Favorable	0.6	0.8	0.5	0.8	0.8	1.0	0.8
Average	0.8	1.0	0.7	1.3	1.0	1.5	1.0
Unfavorable	1.2	1.5	1.1	1.8	1.3	2.0	1.5

(Courtesy TEREX Division, General Motors Corporation.)

Maneuver and dump time is the time required for the scraper to maneuver and dump its load. Average values for maneuver and dump times have also been found to range from about 0.6 to 1.0 min. Table 6-2 provides typical times under varying conditions.

TABLE 6-2 *Scraper maneuver and dump times* (*min*)

Job Conditions	Single-engine, Overhung	3-Axle	All-wheel Drive	Tandem	Elevating
Favorable	0.4	0.4	0.4	0.6	0.4
Average	0.6	0.7	0.6	1.0	0.5
Unfavorable	1.3	1.5	1.0	1.6	0.6

(Courtesy TEREX Division, General Motors Corporation.)

Spot and delay time represents the time required for positioning the scraper at the cut and waiting for pushers to begin loading. This component of scraper cycle time is sometimes neglected, but it must then be compensated for by using an appropriate job efficiency factor. Typical values of spot and delay time are given in Table 6-3.

TABLE 6-3 *Scraper spot and delay times* (*min*)

Job Conditions	Single-engine, Overhung (Push)	3-Axle (Push)	All-wheel Drive		Tandem (Push or Self-load)	Elevating (Self-load)
			Push-load	Self-load		
Favorable	0.4	0.4	0.1	0.1	Negligible	Negligible
Average	0.6	0.6	0.2	0.2	0.1	0.1
Unfavorable	0.8	0.9	0.5	0.3	0.2	0.2

(Courtesy TEREX Division, General Motors Corporation.)

Variable Cycle Time

Variable cycle time is the time spent in hauling the load and returning empty. The maximum speed of the scraper under specific load and hauling conditions can be determined from scraper performance and retarder charts by the methods presented in Chapter 5. To determine travel time, it is necessary to adjust the maximum vehicle speed to compensate for vehicle acceleration and deceleration. Some manufacturers provide travel time curves such as shown in Figures 6-3 and 6-4 for empty and rated load conditions. Such curves include an allowance for acceleration and deceleration. Since there are no travel time curves for negative effective grades, however, speed from the retarder curve must be used to compute travel time in this case.

A second method of compensating for acceleration and deceleration is to apply a speed factor to the maximum vehicle speed to obtain an average vehicle speed. A table of these speed factors is given in Table 6-4. Average speed for a section of scraper route is obtained by multiplying the maximum vehicle speed for the section by the appropriate speed factor from Table 6-4. For the final section of the loaded haul route, use the upper section (under 300 lb/hp) and the column labeled "Level Haul Unit Starting from 0 MPH" to account for the deceleration of the vehicle into the dump area.

Adjustment for Altitude

When a scraper is operating at an altitude above its rated operating level, it is necessary to adjust the unit's power and speed for altitude. When the average speed method is used with performance charts, the maximum speed is adjusted as described in Chapter 5. That is, the required pull (total resistance) is divided by the quantity (1 — derating factor) to give an adjusted value of required pull before entering the performance chart to find maximum speed.

If travel time curves are being used, the travel time obtained from the curve is multiplied by the quantity (1 + derating factor) to obtain an adjusted value of travel time for use in cycle time calculations. In both cases the derating factor must be expressed as a decimal.

Average Payload

The value of scraper payload to be used in production estimation is determined in the same manner as for other haul units. That is, the lesser of the rated weight payload or heaped volume capacity is used. A somewhat different value of average load may be obtained when the optimum load method of Section 6.4 is used. The following example will illustrate the method of estimating scraper production in an earthmoving situation.

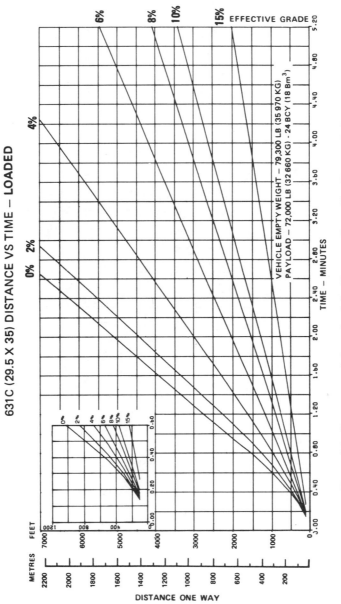

631C (29.5 X 35) DISTANCE VS TIME – LOADED

Figure 6-3   Scraper travel time—loaded. (Courtesy Caterpillar Tractor Co.)

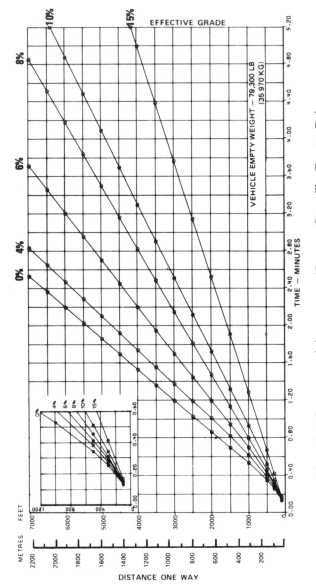

Figure 6-4 Scraper travel time—empty. (Courtesy Caterpillar Tractor Co.)

TABLE 6-4 *Average speed factors*

Haul Road Length in Feet	Level Haul Unit Starting from 0 MPH	Unit in Motion When Entering Haul Road Section		
		Level	Downhill Grade	Uphill Grade Factor
Under 300 lb/hp				
0–200	0–.40	0–.65	0–.67	1.00
201–400	.40–.51	.65–.70	.67–.72	
401–600	.51–.56	.70–.75	.72–.77	(Entrance speed
601–1,000	.56–.67	.75–.81	.77–.83	greater than
1,001–1,500	.67–.75	.81–.88	.83–.90	maximum
1,501–2,000	.75–.80	.88–.91	.90–.93	attainable speed
2,001–2,500	.80–.84	.91–.93	.93–.95	on section)
2,501–3,500	.84–.87	.93–.95	.95–.97	
3,501 and up	.87–.94	.95–	.97–	
300–380 lb/hp				
0–200	0–.39	0–.62	0–.64	1.00
201–400	.39–.48	.62–.67	.64–.68	
401–600	.48–.54	.67–.70	.68–.74	(Entrance speed
601–1,000	.54–.61	.70–.75	.74–.83	greater than
1,001–1,500	.61–.68	.75–.79	.83–.88	maximum
1,501–2,000	.68–.74	.79–.84	.88–.91	attainable speed
2,001–2,500	.74–.78	.84–.87	.91–.93	on section)
2,501–3,500	.78–.84	.87–.90	.93–.95	
3,501 and up	.84–.92	.90–.93	.95–.97	
380 and up lb/hp				
0–200	0–.33	0–.55	0–.56	1.00
200–400	.33–.41	.55–.58	.56–.64	
401–600	.41–.46	.58–.65	.64–.70	(Entrance speed
601–1,000	.46–.53	.65–.75	.70–.78	greater than
1,001–1,500	.53–.59	.75–.77	.78–.84	maximum
1,501–2,000	.59–.62	.77–.83	.84–.88	attainable speed
2,001–2,500	.62–.65	.83–.86	.88–.90	on section)
2,501–3,500	.65–.70	.86–.90	.90–.92	
3,501 and up	.70–.75	.90–.93	.92–.95	

(Courtesy TEREX Division, General Motors Corporation.)

Example 6-1

PROBLEM: Estimate the production of a wheel tractor-scraper given the job information below. Use the travel time curve method with Figures 6-3 and 6-4. Assume that weight governs load capacity.

Scraper type: single-engine overhung.

Scraper weight: 148,700 lb loaded, 76,700 lb empty.

Weight on drivers: 77,400 lb loaded, 51,600 lb empty.

Material: sandy clay, 3,200 lb/BCY, coefficient of traction = 0.60, rolling resistance = 100 lb/ton.

Scraper capacity: rated weight load = 72,000 lb.

Altitude: 7,000 ft.

Job efficiency factor: 0.83.

Haul route:

> Section 1. Level loading area.
> Section 2. Down a 4%-grade, 2,000 ft.
> Section 3. Level dumping area.
> Section 4. Up a 4%-grade, 2,000 ft.
> Section 5. Level turnaround, 600 ft.

Solution:

$$\text{Payload} = \frac{72,000}{3,200} = 22.5 \text{ BCY}.$$

Traction limitation:

> Loaded = 77,400 × 0.60 = 46,440 lb.
> Empty = 51,600 × 0.60 = 30,960 lb.

Resistance:
  Grade:

> Haul = 148,700 × −0.04 = −5,948 lb.
> Return = 76,700 × 0.04 = 3,068 lb.
> Turnaround = 0.

  Rolling:

$$\text{Haul} = \frac{148,700}{2,000} \times 100 = 7,435 \text{ lb.}$$

$$\text{Return and turnaround} = \frac{76,700}{2,000} \times 100 = 3,835 \text{ lb.}$$

Total:

> Haul = 7,435 − 5,948 = 1,487 lb.
> Return = 3,835 + 3,068 = 6,903 lb.
> Turnaround = 0 + 3,835 = 3,835 lb.

Since the minimum traction produced is always greater than the maximum pull required, no traction limitation applies.

Altitude adjustment:

$$\text{Derating factor} = 3 \times \left(\frac{7,000 - 3,000}{1,000} \right) = 12\% \text{ (Equation 5-6).}$$

Variable cycle:

> Section 2 = 1.0 min (Figure 6-3).
> Section 4 = 1.6 min (Figure 6-4).
> Section 5 = 0.35 min (Figure 6-4).
> Adjusted for altitude = (1.0 + 1.6 + 0.35) × 1.12 = 3.30 min.

Fixed cycle:

Load $= 0.8$ min (Table 6-1).
Maneuver and dump $= 0.6$ min (Table 6-2).
Spot and delay $= 0.6$ min (Table 6-3).
Total $= 2.0$ min.

Total cycle time $= 3.3 + 2.0 = 5.3$ min.

Estimated production $= 22.5 \times \dfrac{60}{5.3} \times 0.83 = 211.4$ BCY/hr.

Example 6-2

PROBLEM: Solve the problem of Example 6-1 by using the average speed method and the scraper performance curves of Figures 5-6 and 5-7. Scraper weight is 150,000 lb loaded and 80,000 lb empty. The scraper's weight to power ratio is 361 lb/hp loaded and 193 lb/hp empty.

Solution:

$$\text{Payload} = \frac{150,000 - 80,000}{3,200} = 24.9 \text{ BCY.}$$

Resistance:
  Grade (%):

Haul $= -4.0\%$.
Return $= 4.0\%$.
Turnaround $= 0\%$.
Rolling $= \dfrac{100}{20} = 5.0\%$.

Total (%):

Haul $= 5.0 - 4.0 = 1.0\%$.
Return $= 5.0 + 4.0 = 9.0\%$.
Turnaround $= 5.0 + 0.0 = 5.0\%$.

Read total pull required in pounds from Figure 5.6.
Total pull in pounds adjusted for altitude:

$$\text{Haul} = \frac{1,500}{(1.0 - 0.12)} = \frac{1,500}{0.88} = 1,705 \text{ lb.}$$

$$\text{Return} = \frac{7,200}{0.88} = 8,182 \text{ lb.}$$

$$\text{Turnaround} = \frac{4,000}{0.88} = 4,545 \text{ lb.}$$

Maximum speed $=$ (Figure 5-6):

Section 2 (Haul) $= 32$ mph.
Section 4 (Return) $= 15$ mph.
Section 5 (Turnaround) $= 28$ mph.

Average speed (Table 6-4):

Section 2 (Haul) = $0.80 \times 32 = 25.6$ mph.
Section 4 (Return) = $1.0 \times 15 = 15.0$ mph.
Section 5 (Turnaround) = $0.75 \times 28 = 21.0$ mph.

Variable cycle:

$$\text{Travel time} = \frac{2{,}000}{25.6 \times 88} + \frac{2{,}000}{15.0 \times 88} + \frac{600}{21.0 \times 88} = 2.73 \text{ min.}$$

(1 mph = 88 fpm)

Fixed cycle = 2.0 min (Example 6-1).

Total cycle = $2.0 + 2.73 = 4.73$ min.

$$\text{Estimated production} = 24.9 \times \frac{60}{4.73} \times 0.83 = 262.2 \text{ BCY/hr.}$$

6-3 PUSH-LOADING

General

Wheel tractor scrapers usually require the assistance of pusher tractors to obtain maximum production and lowest cost per cubic yard hauled. Methods of push-loading, the determination of the number of pushers required to serve a scraper fleet, and the estimation of scraper production costs will be covered in this section. The use of push-pull scrapers will also be covered.

Push-Loading

The three basic methods of operation for push tractors are illustrated in Figure 6-5. Back-track loading is the slowest of these loading methods because of the longer travel distance between loading positions. Chain and shuttle loading methods are almost equally efficient. The choice of loading methods will thus depend on the length of the cut area and whether or not scraper traffic can move through the cut area in both directions.

Total pusher cycle time is the sum of the cycle components contained in Equation 6-2.

pusher cycle time = load time + boost time + return time + transfer time (6-2)

Load time is the time during which the scraper is actually loading. Boost time is the time after loading is completed that the pusher remains in contact with the scraper to assist the scraper in accelerating out of the cut area. Return time is the time required for the pusher to return to the start of the loading area. Transfer time

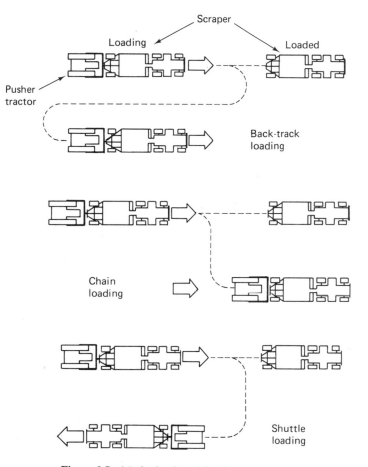

Figure 6-5 Methods of push-loading scrapers.

is the time required for the pusher to move into contact with the next scraper and begin another loading cycle.

Field studies have shown that boost and transfer times average about 0.25 min. Return time for back-track loading is about 0.4 of load time and somewhat less for chain and shuttle loading. Thus, Equation 6-3 may be used for estimating pusher cycle time using the back-track method.

$$\text{pusher cycle time} = 0.25 + (1.4 \times \text{load time}) \tag{6-3}$$

Pusher cycle time for average loading times may also be estimated by multiplying scraper loading time by the factors given in Table 6-5. Tandem pushing utilizes two tractors operating in tandem (one behind the other). As mentioned in Chapter 5, dual tractors are more efficient than tandem tractors because the dual tractor is controlled by a single operator.

TABLE 6-5 *Pusher factors*

Loading Method	Single Pusher	Tandem Pusher
Back-track loading	1.5	2.0
Chain loading or shuttle loading	1.3	1.5

(Courtesy TEREX Division, General Motors Corporation.)

Number of Pushers Required

The number of scrapers that may be fully served (i.e., scrapers do not have to wait for pushers) may be determined by using Equation 6-4. It is suggested that the result be rounded down to one decimal place for use in Equation 6-5. The number of pushers required to serve a fixed number of scrapers may then be found by using Equation 6-5. The result of Equation 6-5 must be rounded up to the next whole number to insure that scrapers do not have to wait for pushers. Additional considerations involved in selecting the number of scrapers to be served by one pusher will be covered in the following section.

$$\text{number of scrapers served} = \frac{\text{scraper cycle time}}{\text{pusher cycle time}} \tag{6-4}$$

$$\text{number of pushers required} = \frac{\text{number of scrapers}}{\text{number of scrapers served by one pusher}} \tag{6-5}$$

Example 6-3

PROBLEM: A wheel scraper has a loading time of 1.1 min and a total cycle time of 6.5 min. Find the number of pushers required to fully serve a fleet of 9 scrapers. Use Table 6-5 with back-track and chain loading methods.

Solution:

Pusher cycle time (Table 6-5):

$$\text{Back track} = 1.5 \times 1.1 = 1.65 \text{ min.}$$
$$\text{Chain} = 1.3 \times 1.1 = 1.43 \text{ min.}$$

Number of scrapers per pusher (Equation 6-4):

$$\text{Back track} = \frac{6.5}{1.65} = 3.9.$$

$$\text{Chain} = \frac{6.5}{1.43} = 4.5.$$

Number of pushers required (Equation 6-5):

$$\text{Back track} = \frac{9}{3.9} = 2.3 = 3.$$

$$\text{Chain} = \frac{9}{4.5} = 2.0 = 2.$$

Calculating Cost Performance

The equipment cost per cubic yard of earth moved is calculated by dividing the hourly equipment cost by the hourly production. Production units used may, of course, be either bank, loose, or compacted cubic yards. Scraper cost performance may be calculated for a single scraper or for the entire fleet. Fleet cost would include not only scrapers but also pushers, graders, spreaders, compactors, and any other equipment required in the cut and fill operation.

Example 6-4

PROBLEM: Determine the scraper cost performance and fleet cost performance for the following equipment fleet operating under the conditions of Example 6-1. The back-track method of loading will be used.

Equipment	Cost ($/hr/item)
Scrapers (6)	36
Pushers (*)	33
Grader (1)	13
Compactor (1)	25

*Use minimum number of pushers required to fully serve scrapers.

Solution:

Pusher cycle time $= 1.5 \times 0.8 = 1.2$ min (Table 6-5).

Maximum number of scrapers/pusher $= \dfrac{5.3}{1.2} = 4.4$ (Equation 6-4).

Number of pushers required $= \dfrac{6}{4.4} = 1.4$; use 2 (Equation 6-5).

Scraper cost performance $= \dfrac{36}{211.4} = \$0.170/\text{BCY}.$

Fleet cost performance $= \dfrac{(6)(36) + (2)(33) + (13) + (25)}{(6)(211.4)} = \$0.252/\text{BCY}.$

Push-Pull Operations

Push-pull is a relatively new form of scraper operation that uses specially designed scrapers capable of coupling up during loading and then uncoupling for hauling. The sequence of operations during loading is as follows:

1. First unit starts to self-load.

2. Second scraper makes contact, couples, and then push-loads front scraper.

3. Front unit pull-loads rear scraper.

4. Units separate for haul to fill.

Some of the advantages claimed for push-pull operations include the following:

1. No pusher and pusher operator required.

2. No possibility of scraper/pusher mismatch.

3. No lost time due to pusher downtime.

4. Lower investment in the equipment fleet.

5. Provides the advantages of self-loading scrapers while retaining the hauling advantages of standard scrapers.

The conditions that favor using push-pull equipment include relatively easy to load materials and long, straight hauls. Since units dump almost simultaneously, there must be an adequate amount of spreading and compacting equipment available. Economic application zones and other considerations in the use of push-pull equipment will be covered in Section 6.5.

6-4 OPTIMUM LOAD TIME

General

Observation of a large number of scraper jobs by the Caterpillar Tractor Co. (see Reference 5 at the end of this chapter) has indicated that most scraper operators always try to obtain a maximum scraper load because they believe that this will yield maximum production. However, field and theoretical studies by Caterpillar show that this assumption is not valid. These studies have led to the development of a method of analysis for determining the most efficient loading time for a particular scraper job. This section will explain the use of this method, commonly called the *optimum load time* method.

Load Growth Curve

In order to utilize the optimum load time method, it is necessary to determine the loading characteristics of a particular combination of soil, scraper, pusher, and job conditions. This can be done by loading the scraper for controlled periods of time, weighing the loads obtained, and plotting the average results. The resulting curve is called a *load growth curve*. A typical load growth curve and plotting data are illustrated in Figure 6-6. The slope of the load growth curve at any loading time represents the rate of loading. The use of different amounts of pusher power for a particular scraper and job combination will yield a family of curves with the most powerful pusher curve lying to the left and above the other curves, as indicated by the dotted curve of Figure 6-6.

Load time (min.)	Load (BCY)
0.2	11.4
0.4	18.7
0.6	22.7
0.8	24.8
1.0	26.1
1.2	26.8
1.4	27.2

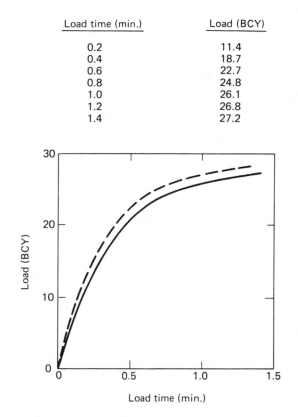

Figure 6-6 Typical load growth curve.

Scraper Optimum Load Time—Numerical Method

Since the loading rate (slope of load growth curve) decreases as loading time increases, there exists some loading time beyond which the scraper production per hour will actually decrease. This point may be found by numerical or graphical methods. Calculation of the optimum load time using data from the load growth curve of Figure 6-6 and a total cycle time minus load time of 3.0 min is illustrated below.

Load time (min)	0.2	0.4	0.6	0.8	1.0	1.2	1.4
Cycle time less load time (min)	3.0	3.0	3.0	3.0	3.0	3.0	3.0
Cycle time (min)	3.2	3.4	3.6	3.8	4.0	4.2	4.4
Trips/50 min	15.6	14.7	13.9	13.2	12.5	11.9	11.4
Load/trip (BCY)	11.4	18.7	22.7	24.8	26.1	26.8	27.2
Production (BCY/hr)	177.8	274.9	315.5	327.4	326.3	318.9	310.1

It can be seen from the results that maximum scraper production is reached with a load time between 0.8 and 1.0 min. The results may be plotted (if desired) to identify optimum load times that fall between the calculated points.

Scraper Optimum Load Time—Graphical Method

The optimum load time for maximum scraper production may also be found rather easily by a graphical method. The procedure for using this method is as follows. First, extend the horizontal axis of the load growth curve to the left, as shown in Figure 6-7. Next, locate a point (point *A*) on this axis which lies a distance equal to the total cycle time minus load time to the left of zero load time. Then, draw a straight line from this point tangent to the load growth curve. Finally, drop a vertical line from the point of tangency (point *B*) to the horizontal axis (point *C*). The load time corresponding to point *C* is then the optimum load time. Again, the optimum load time is found to be about 0.9 min.

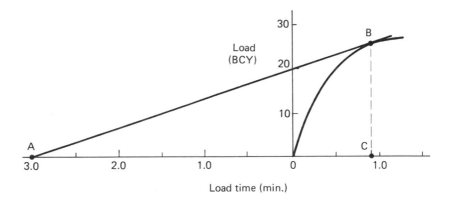

Figure 6-7 Graphical method for finding scraper optimum load time.

An analysis of Figure 6-7 will prove that point *C* is in fact the optimum load time. The distance *A–C* represents the total cycle time for this particular load time. The distance *B–C* represents the resulting production in bank cubic yards per load. Since the horizonal scale is in minutes, the slope of the line *A–B* represents the production in bank cubic yards per minute. The maximum slope of the line connecting total cycle time and load time occurs at the point of tangency (point *B*). Hence, point *C* represents the load time that will yield the maximum production per unit of time with other factors fixed.

Maximizing Pusher Production

The preceding determination of optimum load time was based on maximizing scraper production and assumed that an adequate number of pushers were available to insure that scrapers did not have to wait for pushers. When pusher resources are limited, an optimum load time based on optimizing pusher production rather than scraper production will be required.

Pusher optimum load time may be found by using similar numerical or graphical methods. However, as noted earlier, pusher cycle time is itself a function of loading time. Thus, when the graphical method (Figure 6-8) is used, a loading time must first be assumed. The corresponding pusher cycle time is then calculated, plotted (point A_1) and a tangent drawn to the load growth curve (point B). The optimum load time found (point C) is then compared with the assumed load time. If there is an appreciable difference, then a new pusher cycle time based on the load time of point C is calculated and plotted (point A_2). The process is repeated until the optimum load time from the graph is approximately equal to the load time used for calculating pusher cycle time. Usually, only one or two cycles are required to obtain a satisfactory result.

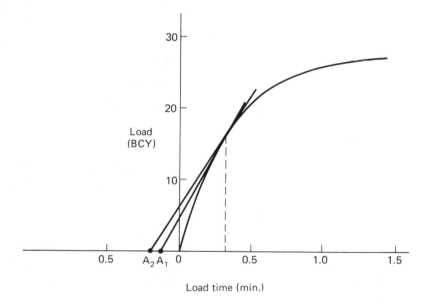

Figure 6-8 Determining pusher optimum load time.

The numerical method of determining pusher optimum load time is illustrated by the calculations below. Pusher cycle time is calculated by using Equation 6-3.

Load time (min)	0.2	0.4	0.6	0.8	1.0	1.2	1.4
Pusher cycle	0.53	0.81	1.09	1.37	1.65	1.93	2.21
Cycles/50 min	94.3	61.7	45.9	36.5	30.3	25.9	22.6
Load/cycle (BCY)	11.4	18.7	22.7	24.8	26.1	26.8	27.2
Production (BCY/hr)	1,075.0	1,153.8	1,041.9	905.2	790.8	694.1	614.7

The maximum production that can be attained by using one pusher may be shown by plotting the points calculated above. This is illustrated by Figure 6-9.

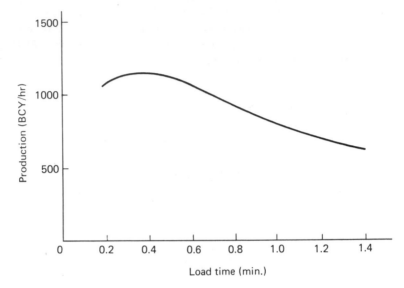

Figure 6-9   Maximum pusher production vs. loading time.

Maximizing Fleet Production

The curves representing production versus loading time for various numbers of scrapers (assuming adequate pusher support) are merely multiples of the curve for a single scraper. This assumes that there will be no interference between scrapers during loading, hauling, or dumping. Thus, Figure 6-10 represents the production of one to five scrapers operating under the conditions of Figure 6-6.

If the pusher production curve of Figure 6-9 were superimposed on Figure 6-10, as shown in Figure 6-11, only those values of production which fall below the shaded area are possible. Maximum production is limited by scraper or pusher production, whichever is lower. In Figure 6-11 it can be seen that the maximum production using one pusher under these conditions is the same as maximum pusher production and amounts to approximately 1,160 BCY/hr. This production is obtained with five scrapers and a load time of about 0.35 min. Note, however, that this may not represent the most efficient use of equipment in terms of cost per cubic yard of earth moved.

Analyzing Production and Cost

The production data used to develop Figure 6-11 may now be used to determine cost performance. Using hourly costs of $36 for each scraper and $33 for

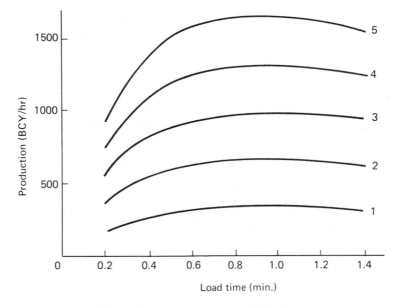

Figure 6-10 Multiple scraper production.

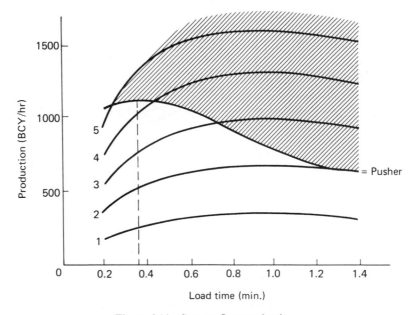

Figure 6-11 Scraper fleet production.

each pusher, cost performance of the fleet is calculated as follows:

Load time (min)	0.2	0.4	0.6	0.8	1.0	1.22	1.4
Production (BCY/hr):							
1 scraper	177.8	274.9	315.5	327.4	326.3	318.9	305.5
2 scrapers	355.6	549.8	631.0	654.8	652.6	637.8	614.7
3 scrapers	533.4	824.7	946.5	905.2	790.8	694.1	614.7
4 scrapers	711.2	1099.6	1041.9	905.2	790.8	694.1	614.7
5 scrapers	889.0	1153.8	1041.9	905.2	790.8	694.1	614.7
Cost performance ($/BCY):							
1 scraper	0.388	0.251	0.219	0.211	0.211	0.216	0.226
2 scrapers	0.295	0.191	0.166	0.160	0.161	0.165	0.171
3 scrapers	0.264	0.171	0.149	0.156	0.178	0.203	0.229
4 scrapers	0.249	0.161	0.170	0.196	0.224	0.255	0.288
5 scrapers	0.240	0.185	0.204	0.235	0.269	0.307	0.347

These data may then be plotted to give fleet production cost per cubic yard as shown in Figure 6-12.

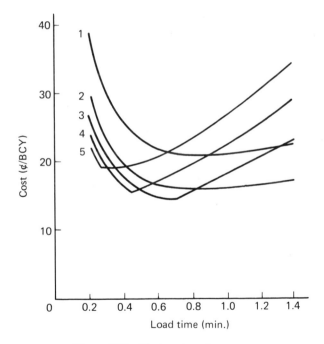

Figure 6-12 Fleet cost performance.

A study of Figure 6-12 indicates that the lowest cost per bank cubic yard of excavation is obtained with three scrapers using a loading time of 0.7 min. Excavation cost will be $0.146/BCY with an output of 969 BCY/hr. Compare this with a maximum production of 1,160 BCY/hr at a cost of $0.184/cu yd using five scrapers.

Since the cost per cubic yard at optimum load time is only slightly higher when using four scrapers, the job planner might choose to use four scrapers to take advantage of their increased production. In addition, with four scrapers if a scraper were lost because of equipment failure, there would actually be a slight decrease in unit cost of excavation instead of an increase.

Further analysis of the cost curve for three scrapers gives additional insight into cost performance at times other than optimum. Figure 6-13 shows fleet cost performance versus loading time for three scrapers. Load times to the left of optimum represent excess pusher capability or pushers waiting for scrapers. Load times to the right of optimum represent excess scraper capacity or scrapers waiting for pushers. Note that the slope of the cost curve is much greater when loading time is too long than when loading time is too short. On the job this means that the cost penalty for loading too long is much greater than the cost penalty for loading too short. This confirms the old rule that scrapers should never be kept waiting for pushers.

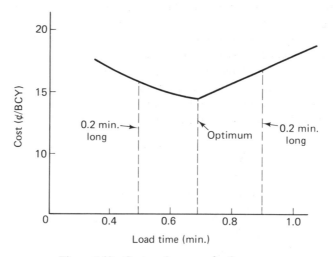

Figure 6-13   Cost performance for 3 scrapers.

The cost for a 0.2-min deviation from optimum load time may be calculated for three scrapers and one pusher as follows:

Load Time (min)	Production (BCY/hr)	Cost Performance ($/BCY)
0.5	900.0	0.157
0.7	968.9	0.146
0.9	847.7	0.166

Thus, while loading 0.2 min short increases cost 7.5% above optimum, loading 0.2 min too long increases cost 13.7% above optimum. Hence, the cost penalty in this case is almost twice as great for loading too long as for loading too short. Other considerations in managing scraper loading operations will be covered in the next section.

6-5 SCRAPER OPERATIONS

Application Zones

As mentioned earlier, scrapers are used primarily on medium to long hauls. The operating area of medium hauls is sometimes referred to as the *slow-speed zone*. This zone is characterized by a haul distance beyond economical bulldozing range but too short to permit high-speed hauling and a relatively high total resistance. Because of these characteristics, either crawler-powered scrapers, all-wheel drive scrapers, or push-pull scrapers might be used in this zone.

Long hauls correspond to the high-speed hauling zone. In this zone will normally be found good ground conditions, well-maintained haul roads, and adequate pusher power. These factors enable the wheel scraper to exploit its high-speed capability. Whenever high total resistance is encountered, tandem-engine scrapers will usually be economical.

Economic application zones as a function of one-way haul distance and total resistance are graphically illustrated in Figure 6-14. In general, single-engine, overhung scrapers operate best on relatively flat haul roads up to about 6,000 ft in length where maneuverability is important and adequate pusher power is

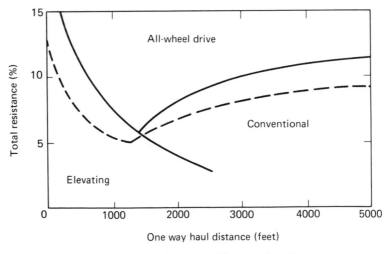

Note: Zone above dashed line = push-pull zone

Figure 6-14 Graphical illustration of economic application zones.

available. Three-axle scrapers are best for long, relatively flat high-speed hauls. All-wheel drive, tandem-powered, and push-pull scrapers can usually load without pushers and can operate under conditions of high total resistance. Elevating scrapers can also self-load and can provide a finish grade, but they operate best on relatively short hauls. A chart providing additional guidance in scraper selection appears as Figure 6-15.

Production Techniques

Some of the techniques that may be used to increase scraper production and reduce cost are discussed below. Additional considerations in job supervision and management will be covered in the next subsection.

Loading

1. Use downhill loading whenever possible. This reduces total resistance and the amount of pusher power required.

2. Shuttle or chain load if possible.

3. Straddle loading may be used to speed up loading. In this method, cuts are spaced at intervals so that a ridge is left between cuts. The material in the ridge can then be loaded faster and easier than can normally cut material.

4. The scraper should leave the cut as fast as possible, spreading the soil left in front of the bowl as the bowl is lifted.

5. Adjust loading techniques according to the soil type encountered. Use rippers or scarifiers to loosen hard soil before loading. Normally, the operator should maintain a depth of cut that gives a constant loading speed and leaves a smooth floor. However, pumping the bowl up and down during loading may be required when loose sand is being loaded.

6. Pushers should provide an adequate boost to accelerate scrapers leaving the cut.

7. Maintain a loading time as close to optimum as possible and provide adequate pushers. Tandem or dual pushers may be required for adequate power.

8. Maintain the cut area with idle pushers or provide full-time maintenance equipment. If bunching occurs in the load area, shorten loading time until the scrapers are again spaced out properly.

Hauling

1. Maintain the haul road in top condition; use graders and water sprinklers as required.

2. Keep haul units uniformly spaced. Unless passing room is provided, hauling speed is limited to that of the slowest unit.

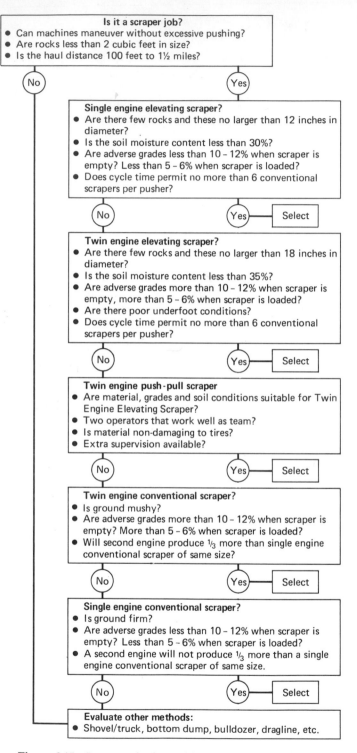

Is it a scraper job?
- Can machines maneuver without excessive pushing?
- Are rocks less than 2 cubic feet in size?
- Is the haul distance 100 feet to 1½ miles?

No / Yes

Single engine elevating scraper?
- Are there few rocks and these no larger than 12 inches in diameter?
- Is the soil moisture content less than 30%?
- Are adverse grades less than 10 – 12% when scraper is empty? Less than 5 – 6% when scraper is loaded?
- Does cycle time permit no more than 6 conventional scrapers per pusher?

No / Yes — Select

Twin engine elevating scraper?
- Are there few rocks and these no larger than 18 inches in diameter?
- Is the soil moisture content less than 35%?
- Are adverse grades more than 10 – 12% when scraper is empty, more than 5 – 6% when scraper is loaded?
- Are there poor underfoot conditions?
- Does cycle time permit no more than 6 conventional scrapers per pusher?

No / Yes — Select

Twin engine push-pull scraper
- Are material, grades and soil conditions suitable for Twin Engine Elevating Scraper?
- Two operators that work well as team?
- Is material non-damaging to tires?
- Extra supervision available?

No / Yes — Select

Twin engine conventional scraper?
- Is ground mushy?
- Are adverse grades more than 10 – 12% when scraper is empty? More than 5 – 6% when scraper is loaded?
- Will second engine produce ⅓ more than single engine conventional scraper of same size?

No / Yes — Select

Single engine conventional scraper?
- Is ground firm?
- Are adverse grades less than 10 – 12% when scraper is empty? Less than 5 – 6% when scraper is loaded?
- A second engine will not produce ⅓ more than a single engine conventional scraper of same size.

No / Yes — Select

Evaluate other methods:
- Shovel/truck, bottom dump, bulldozer, dragline, etc.

Figure 6-15 Scraper selection guide. (Courtesy WABCO Construction and Mining Equipment Group.)

3. Use one-way haul roads if possible. When two-way traffic is necessary, provide grade separation on sharp curves to reduce the danger of collision.

4. Haul from two cuts to a common fill or haul from one cut to two fills whenever possible. Either procedure reduces time lost in turning which typically amounts to about 0.25 min per turn.

5. Use sprinklers to reduce dust on the haul road. Dust increases equipment wear and reduces operator visability. Do not overwater because over-watering may cause loss of traction and be a greater hazard than dust.

Spreading

1. Spread the first load at the beginning of each fill lane and work down the fill so that the loaded scrapers compact the material spread previously.

2. Keep the fill high on the outside for safety and to help maintain an accurate slope. However, drainage must be provided for the fill surface.

3. Provide adequate spreading and compacting equipment and give scrapers the right-of-way on the fill.

4. Vary spreading procedures according to the material involved. Spread loose sand in thin layers. Wet and sticky material is best spread by moving the tailgate forward about 12 in. and then returning it about 6 in. before repeating the procedure. When using this procedure, be sure to keep the bowl high enough to allow the dumped material to pass under the scraper without clogging.

5. Keep the fill smooth enough to allow fast dumping and return to the haul road.

6. Use boosting on the fill if scrapers require extra power for fast, smooth spreading.

Job Management

Planners and supervisors must insure that operations actually proceed in the manner intended and that adjustments are made as conditions change. Use well-qualified personnel to control operations in the cut, on the haul, and in the fill. Operators must know what is wanted. The cut foreman or pusher operator can signal the scraper operator when an optimum load is obtained. Seconds lost in several phases of the operation add up rapidly to increase total cycle time and reduce production. Sixty seconds lost on what should be a 5 min total cycle will reduce production and increase cost by 20%.

Units on the haul road must be operated near their rated revolutions per minute and not allowed to lug down. Keep haul units evenly spaced except in push-pull operations where they move in pairs. Units hauling too slowly should be speeded up by light loading or pulled off for repair. Operators hauling at excessive speeds must be corrected. Use flagmen at blind spots and at highway and railroad crossings.

Develop a sound maintenance program and provide for emergency on-site repairs. Maintenance and repair will be covered in detail in Chapter 13. Have a plan for continuing operations whenever a major piece of equipment breaks down.

PROBLEMS

1. Estimate the fixed cycle time for a single-engine, overhung scraper under the following conditions:
   (a) Average conditions.
   (b) Poor conditions.
   (c) Favorable conditions.

2. If the haul and return time for the scraper in Problem 1 is 4.0 min, find the difference in the number of cycles per hour that could be completed with favorable conditions and with poor conditions. Use a job efficiency factor of 0.83.

3. Find the number of pushers required to serve a fleet of six scrapers whose cycle time is 5.0 min and whose loading time is 1.0 min. The back-track method of loading will be used. Find pusher cycle time by using both Equation 6-3 and Table 6-5 (single pusher).

4. The scraper whose load growth curve is shown in Figure 6-6 has a cycle time minus load time of 4.0 min. Find the scraper's optimum loading time by using the graphical method.

5. Using Equation 6-3 for pusher cycle time, find the optimum pusher loading time in the situation in Problem 4 by using the graphical method.

6. Use the numerical method to find scraper production versus loading time for the situation of Problem 4. Plot the results for one to five scrapers. Use a 50-min hour.

7. Use the numerical method to find pusher production versus loading time for the situation in Problem 4. Plot the result and superimpose on the graph of Problem 6.

8. The load growth data for a scraper are given below. The scraper's total cycle time minus load time is 4.0 min and the pusher's cycle time minus load time is 0.65 min. Equipment owning and operating costs are $35/hr per scraper and $30/hr for the pusher. Use a job efficiency factor of 0.83.
   (a) What is the minimum cost per bank cubic yard of production for a fleet of these scrapers that uses one pusher?
   (b) What is the optimum load time?
   (c) What is the hourly production of the fleet?
   (d) How many scrapers are required to obtain this production?

Loading Data	
Load Time (min)	Average Load (BCY)
0.2	11.0
0.4	18.0
0.6	21.8
0.8	23.9
1.0	25.1
1.2	25.9

9. You are given the following data on a scraper job. Six single-engine, overhung scrapers (whose travel time curves are shown in Figures 6-3 and 6-4) and one pusher will be used along with a dozer for spreading the fill and a grader for maintaining the haul route. Equipment owning and operating costs are $30/hr per scraper, $25/hr per dozer, and $12/hr for the grader. Assume that the scraper will carry 24 BCY. A single-haul route will be used for haul and return. Efficiency factor is 0.83 and job conditions are average. Use Equation 6-3 for pusher cycle time. Sections of the haul route from the cut to the fill are as follows:

Section	Distance (ft)	Grade (%)	Rolling Resistance (lb/ton)
1	1,200	−3.0	100
2	4,000	+1.0	150
3	500	0	200

(a) What is the estimated fleet production in bank cubic yards per hour?
(b) What is the fleet cost per bank cubic yard produced?
(c) Is one pusher adequate for this situation?

10. Use the average speed method and Figures 5-6 and 5-7 to find the haul and return travel time for the job of Problem 9. The scraper has a 415-hp engine.

REFERENCES

1. *Basic Estimating* (3rd ed.). Melrose Park, Illinois: Construction Equipment Division, International Harvester Company, n.d.

2. *Caterpillar Performance Handbook*. Peoria, Illinois: Caterpillar Tractor Co., 1975.

3. CORRIGAN, FRANK W., "Making Scrapers Pay," *Construction Methods and Equipment*, Vol. 49, No. 2, (February 1967), pp. 90–94, 101–102.

4. *Economics of Large Scrapers*. Peoria, Illinois: Caterpillar Tractor Co., n.d.

5. *Optimum Load Time*. Peoria, Illinois: Caterpillar Tractor Co., 1968.

6. *Production and Cost Estimating of Material Movement with Earthmoving Equipment*. Hudson, Ohio: TEREX Division, General Motors Corporation, 1970.

7

COMPACTION

7-1 THE COMPACTION PROCESS

General

Compaction is the process of artificially increasing the density of soils by forcing the soil particles closer together, primarily by expelling air from the void spaces in the soil. An increase in soil density caused by the expulsion of water from the void spaces is referred to as *consolidation*. Consolidation is a long-term process normally requiring months or years while compaction takes place in a much shorter time.

The increased density obtained by soil compaction improves the construction characteristics of the soil in several respects. Among these improved characteristics are the following:

1. Increased strength

2. Reduced compressibility

3. Improved volume change characteristics

4. Reduced permeability

The amount of compaction that can be achieved with a given soil depends on the soil's initial density, its physical and chemical characteristics (grain-size distribution, cohesiveness, etc.), moisture content, and amount and type of compactive effort applied.

In order to properly select and utilize compaction equipment, it is necessary to understand how soil density is specified and measured for construction purposes. The following paragraphs will discuss these topics as well as the nature of compaction forces.

Required Density and Optimum Moisture Content

The soil density required for a construction project is usually specified as a percentage of the maximum density obtained for that soil in a standard laboratory test. The laboratory test most commonly used is an impact compaction test often referred to as a *Proctor test* or an *AASHO* (American Association of State Highway Officials) *test*. Actually, there are two forms of this test: the Standard Proctor or Standard AASHO (T-99) and the Modified Proctor or Modified AASHO (T-180). Characteristics of the tests are given in Table 7-1. Note that the compactive effort for the modified test is more than four times as great as for the standard test and thus yields a greater soil density. Hence, the modified test is used whenever high design loads will be encountered, for example, for airport runways and taxiways.

TABLE 7-1 *Characteristics of AASHO compaction tests*

Test Details	Standard	Modified
Diameter of mold (in.)	6	6
Height of sample (in.)	5 cut to 4.59	5 cut to 4.59
Numbers of layers	3	5
Blows per layer	25	25
Weight of hammer (lb)	$5\frac{1}{2}$	10
Diameter of hammer (in.)	2	2
Height of hammer drop (in.)	12	18
Volume of sample (cu ft)	$\frac{1}{30}$	$\frac{1}{30}$
Compactive effort (ft-lb)	12,400	56,200

To determine the maximum density of a soil using AASHO test methods, compaction tests are run over a range of moisture contents. The results are then plotted as dry density versus moisture content as illustrated in the Standard AASHO curve of Figure 7-1. The peak of this curve (about 105 lb/cu ft dry density) represents the maximum density obtained under test conditions. Hence, 100% of Standard AASHO density corresponds to a dry density of 105 lb/cu ft, 90% Standard AASHO corresponds to 94.5 lb/cu ft, etc.

The moisture content at which maximum density is obtained is referred to as the *optimum moisture content* of the soil. Referring to Figure 7-1, we see that the optimum moisture content of this soil using the Standard AASHO test is about 20% of the soil's dry weight. For the Modified AASHO test, however, the optimum

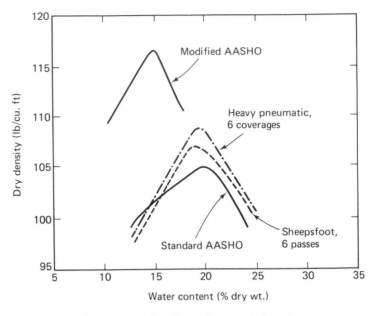

Figure 7-1 Soil moisture-density relationship.

moisture content decreases to 15% while the maximum dry density increases to 117 lb/cu ft. Thus, for a typical soil, the optimum moisture content decreases and the maximum dry density increases with an increase in compactive effort.

The compactive effort obtained under field conditions will not normally correspond exactly to either the Standard or Modified AASHO test. However, a plot of density versus moisture content will usually display a similar characteristic shape. Results of two different field compaction efforts on this soil are also illustrated in Figure 7-1. Comparison of field and laboratory results for different soil types using various types of compactive equipment and varying compactive effort will be discussed in the following section.

Measuring Field Density

In order to obtain proper construction quality control, it is necessary to determine soil density actually obtained in the field and compare this with the specified soil density. There are a number of accepted methods for determining in-place soil density. Some of these will be briefly described below. Note that all of these test methods, except the use of the nuclear density device, involve removing a soil sample and then making an accurate determination of the in-place sample volume (by measuring the volume of the hole produced) and sample dry weight.

Liquid tests measure the volume of the sample hole by measuring the volume of liquid required to fill the hole. For relatively impermeable soils, this test may be performed by pouring a viscous liquid, such as SAE 40 lubricating oil, directly into the hole from a calibrated container. Another method involves forcing water from a calibrated container into a rubber balloon placed in the hole.

Sand tests use a very uniform sand to fill both the hole and a calibrated funnel which is placed over the hole. The sample volume is then found as the volume of sand used less the volume required to fill the funnel. In recent years nuclear density devices capable of measuring both soil moisture and density have come into use. When properly calibrated and operated, they show excellent correlation with the standard test methods previously discussed. Their main advantage lies in their ability to measure dry density in a fraction of the time required by conventional methods.

Compactive Forces

There are four principal forces involved in compaction: static weight, manipulation, impact, and vibration. All compaction equipment utilizies static weight or pressure to achieve compaction. Most compactors combine static weight with one or more of the other compaction forces. Since soil tends to be displaced laterally by the force of the compaction effort, the most efficient compaction is obtained when such displacement is minimized. Manipulation or kneading of soil while under pressure assists in achieving compaction in many soils, particularly the plastic soils. Impact and vibration are also helpful in compaction. The forces involved are similar except for their frequency. Impact involves delivering blows at a low frequency, usually below 10 cycles per second. Vibration involves higher frequencies and may extend to 80 or more cycles per second. Vibration is particularly effective in compacting cohesionless soils such as sand.

7-2 COMPACTION EQUIPMENT

Types of Equipment

Following are the principal types of compaction equipment used in construction. They are illustrated in Figure 7-2. Some compactors are actually a combination of the following:

1. Tamping foot rollers

2. Grid or mesh rollers

3. Vibratory compactors

4. Smooth steel drums

5. Pneumatic rollers

6. Segmented pad rollers

Smooth, steel wheel roller.

Self-propelled vibrating roller.

Small, mutli-tired pneumatic roller.

Heavy pneumatic roller.

Self-propelled tamping foot roller.

Self-propelled segmented steel wheel roller.

Towed sheepsfoot roller.

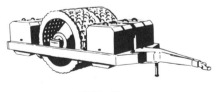

Grid roller.

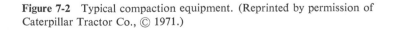

Figure 7-2 Typical compaction equipment. (Reprinted by permission of
Caterpillar Tractor Co., © 1971.)

Tamping Foot Rollers

Tamping foot rollers use a drum that is equipped with a number of protruding feet to achieve compaction. These rollers come with a variety of foot shapes and sizes and include the classic sheepsfoot roller. The sheepsfoot roller received its name from the fact that its tamping action resembles that of a flock of sheep. It is designed to achieve compaction through a combination of static weight (pressure) and manipulation. It also produces some impact force but tends to displace and tear the soil when entering and leaving the soil surface. Sheepsfoot rollers are usually used with multiple drums and may be towed in tandem. The amount of contact pressure varies with the contact area of the feet and the weight of the roller.

Other types of tamping foot rollers use a somewhat different foot design than do sheepsfoot rollers. Generally, a tapered shank is used with a foot designed to minimize lateral displacement of soil during entry and withdrawal. High-speed tamping foot rollers deliver impacts at a frequency approaching vibration. Hence, they produce all four types of compactive effort. Such rollers may operate at speeds of 10 mph or more.

Grid or Mesh Rollers

The grid or mesh roller can also operate at relatively high speeds, since it does not tend to scatter soil. It is well-suited to breaking up lumps of cohesive material. It can also be used to both crush and compact soft rocks having a loss of 20% or more in the Los Angeles Abrasion Test. Its compactive effort is due to static weight and impact, with some manipulation.

Vibratory Compactors

Vibratory compactors are available in a wide range of sizes and types from small hand-operated, plate-type compactors (Figure 7-3) to large, self-propelled compactors that use smooth drum (Figure 7-4), tamping foot (Figure 7-5), or segmented pad rollers. Many vibratory compactors permit varying the vibration frequency and amplitude to obtain the most effective compaction. Vibratory compactors tend to dry out the soil, which is an asset when working with moist soils. Compactive forces are principally vibration and static weight.

Smooth Steel Drums

Steel-wheeled rollers are still widely used for compaction and finish work on base courses and bituminous pavements. Typical models include the three-wheel (two-axle) roller, two-axle tandem roller, and three-axle tandem roller. Static weight is the primary compactive force involved.

Figure 7-3 Walk-behind vibratory plate compactor. (Courtesy Wacker Corp.)

Pneumatic Rollers

Pneumatic rollers are available in several types. The principal types are the multi-tired roller and the heavy pneumatic roller. Multi-tired rollers frequently use a wobble-wheel design to impart a kneading action to the soil. Heavy pneumatic rollers are used for compacting thick soil layers to high density. Multi-tired rollers are used principally for finish work on soil and bituminous surfaces. The compactive forces present in pneumatic rollers are primarily static weight and manipulation.

Segmented Pad Rollers

Segmented pad rollers are similar to sheepsfoot and tamping foot rollers except that segmented pads instead of feet are mounted on the drums. Segmented pad rollers are able to compact soil with less surface disturbance than tamping foot rollers. They use all four of the compactive forces.

Figure 7-4 Large smooth drum vibratory roller. (Courtesy Road Division, Koehring Co.)

Figure 7-5 Vibratory tamping foot compactor. (Courtesy RayGo, Inc.)

7-3 SELECTION AND USE OF EQUIPMENT

Selection of Equipment

When a compaction operation is being planned, it is first necessary to select compaction equipment that is most likely to yield the desired density in minimum time and at a reasonable cost. This initial choice of equipment will be governed primarily by the type of soil involved and the expected operating conditions. Table 7-2 provides a guide to suitable compaction equipment and densities likely to be obtained in the Modified AASHO test for each of the soil types. Table 7-3 summarizes the results of a number of field tests on four types of soil using several different types of compaction equipment.

TABLE 7-2 *Soil compaction guide*

Soil Type	*Compaction Equipment* *Recommended*	*Suitable*	*Maximum Dry Density (pcf) Modified AASHO*
GW	VR, VP	PH, SW, SP, GR, CT	125–140
GP	VR, VP	PH. SW, SP, GR, CT	110–140
GM	VR, PH, SP	VP, SW, GR, CT	115–145
GC	PH, SP	SW, VR, VP, TF, GR, CT	130–145
SW	VR, VP	PH, SW, SP, GR, CT	110–130
SP	VR, VP	PH, SW, SP, GR, CT	105–135
SM	VR, PH, SP	VP, SW, GR, CT	100–135
SC	PH, SP	SW, VR, VP, TF, GR, CT	100–135
ML	PH, SP	TF, SW, VR, VP, GR, CT	90–130
CL	PH, SP	TF, SW, VR, GR, CT	90–130
OL	PH, SP	TF, SW, VR, GR, CT	90–105
MH	PH, SP	TF, SW, VR, GR, CT	80–105
CH	TF, PH, SP	VR, GR, SW	90–115
OH	TF, PH, SP	VR, GR, SW	80–110
Pt	Compaction not practical		

Symbols:
  CT = Crawler Tractor 10–30T SW = Smooth Wheel 3–15T
  GR = Grid Roller 5–15T TF = Tamping Foot 5–30T
  PH = Pneumatic Roller 10–50T VP = Vibrating Plate < 1T
  SP = Segmented Pad 5–30T VR = Vibrating Roller 3–25T

Compactor Operations

After an initial selection of equipment has been made, it is necessary to determine the operational procedure to be used for compaction. This will involve determination of the optimum moisture content of the soil for the specific compactor used and action which may be necessary to change the soil's present moisture content. Other variables to be considered include lift thickness (thickness of each layer to be compacted), compactor speed, contact pressure and total weight, and num-

TABLE 7-3 *Compaction test results*

		Heavy Clay		Sandy Clay		Well-graded Sand		Gravel-Sand-Clay	
Type of Soil Compactor		Degree of Compaction Modified AASHO (%)	Optimum Moisture Content* (%)	Degree of Compaction Modified AASHO (%)	Optimum Moisture Content (%)	Degree of Compaction Modified AASHO (%)	Optimum Moisture Content (%)	Degree of Compaction Modified AASHO (%)	Optimum Moisture Content (%)
Pneumatic-tired roller	12 ton	87	23	88	18	98	10	96	8
Ditto	20 ton	91	21	93	15	98	9	101	6
Ditto	45 ton	96	19	95	14	101	9	101	6
Smooth-wheel roller	2.8 ton	83	21	89	17	101	9	99	7
Ditto	8 ton	90	20	92	15	102	9	100	7
Sheepsfoot roller	5 ton	92	16	94	12	—	—	94	6
Ditto	4.5 ton	92	15	94	13	—	—	93	5
Tamper	100 kg	95	18	94	11	99	8	99	6
Ditto	600 kg	91	17	90	14	98	10	99	7
Vibrating roller	200 kg	—	—	—	—	95	11	89	8
Ditto	350 kg	79	28	79	16	98	9	96	8
Ditto	2.5 ton	83	21	—	—	102	7	101	6
Ditto	3.8 ton	91	21	94	14	105	7	105	6
Vibrating plate compactor	200 kg	—	—	—	—	98	10	92	9
Ditto	650 kg	89	21	92	15	104	8	102	6
Ditto	700 kg	75	20	90	16	100	9	99	7
Ditto	1.5 ton	—	—	—	—	99	9	98	8
Ditto	2 ton	84	17	—	—	98	9	99	7
Caterpillar tractor	40 hp	83	22	—	—	98	10	93	8
Ditto	80 bp	85	24	—	—	—	—	91	8

*Optimum moisture content in compaction with the compactor in question.
(Road Research Laboratory) (Forssblad[5])

ber of passes. For vibratory compactors, the frequency and amplitude of vibration must also be considered.

Figure 7-6 illustrates the variation of field optimum moisture content for four different rollers in only one type of soil—a silty clay (CL). Thus, it is apparent that the optimum moisture content for compaction depends not only on the type of soil but also on the type and weight of compaction equipment used. In general, for plastic soils, the optimum moisture content for pneumatic rollers appears to be close to laboratory optimum. However, for tamping foot rollers, the field optimum appears to be appreciably lower than the laboratory optimum. For nonplastic soils, the optimum field moisture appears to average about 80% of laboratory optimum for all types of rollers. When field moisture content is much less than laboratory optimum, the vibratory roller appears to be the most effective compactor, regardless of soil type.

The thickness of each lift is another major variable in compaction operations. The lift thickness which may be used to achieve a required density will vary with

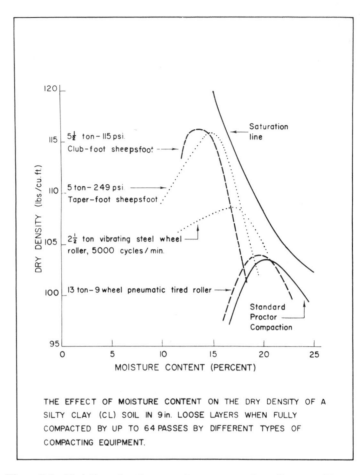

Figure 7-6 Variation of optimum moisture content by roller type. (Townsend[10].)

the soil and compactor characterisics. In general, it is recommended that lifts be kept thin for best results. For tamping foot rollers, the compacted lift thickness should not exceed the length of the roller feet. This suggests a maximum loose lift thickness of 2 in. greater than the length of the feet. For smooth-wheel rollers, loose lift thickness should be limited to about 6 in. Small pneumatic rollers also require relatively thin lifts. For these rollers, a maximum compacted lift thickness of approximately 6 in. is suggested. Heavy pneumatic rollers may achieve satisfactory results with compacted lift thicknesses of 12 in. or more. However, if a compacted lift thickness greater than 12 in. is to be used, it is recommended that the lift be precompacted by a lighter roller before the heavy roller is used. Excessive rutting with a heavy pneumatic roller indicates that the lift thickness is too great, the soil is too wet, or the tire contact pressure is too high. Vibratory compactors are capable of compacting cohesionless soils having lift thickness much greater than could be used with nonvibratory compactors. Lift thicknesses up to 7 ft in depth have been successfully used in compacting rock with heavy vibratory rollers.

Figure 7-7 shows the effect of the number of passes on the soil density for three different rollers on a sandy clay soil. Figure 7-8 shows the variation of density with number of passes for a tamping foot roller on four different types of soil. From these figures it can be seen that the increase in density for most soils and rollers is rather small after about 20 passes.

Additional considerations in the use of specific compactors are given in the following paragraphs.

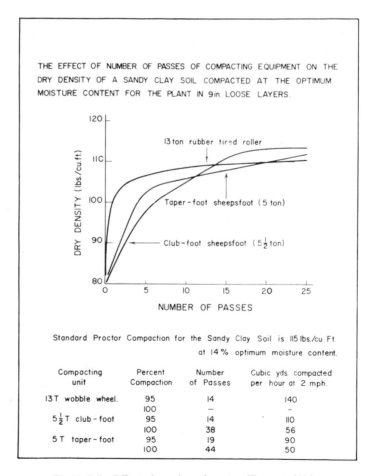

THE EFFECT OF NUMBER OF PASSES OF COMPACTING EQUIPMENT ON THE DRY DENSITY OF A SANDY CLAY SOIL COMPACTED AT THE OPTIMUM MOISTURE CONTENT FOR THE PLANT IN 9 in. LOOSE LAYERS.

Standard Proctor Compaction for the Sandy Clay Soil is 115 lbs./cu Ft. at 14% optimum moisture content.

Compacting unit	Percent Compaction	Number of Passes	Cubic yds. compacted per hour at 2 mph.
13 T wobble wheel.	95	14	140
	100	–	–
5½ T club-foot	95	14	110
	100	38	56
5 T taper-foot	95	19	90
	100	44	50

Figure 7-7 Effect of number of passes. (Townsend[10].)

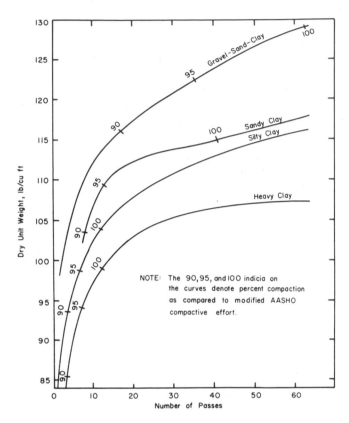

Figure 7-8   Number of passes vs. soil type. (U.S. Department of the Army.)

Tamping Foot Rollers

Tests have indicated that for the range of contact unit pressures usually available, the roller's contact pressure has an insignificant effect on resulting density. However, an increase in foot size (with contact pressure and number of passes remaining constant) will increase soil density. The amount of ground area covered per pass will also increase with increasing foot size. Thus, foot size should be as large as possible while developing the necessary minimum contact pressure. Tests also indicate little change in density due to change in rolling speed as long as tearing and displacement of soil do not become excessive. High-speed tamping foot rollers and segmented pad rollers minimize such displacement. Tamping foot rollers rise to the surface or "walk out" of soil as the soil becomes compacted. After three or four passes, the roller should walk out to within about 1 in. of the surface. If a roller fails to walk out after a reasonable number of passes, it indicates that excessive shearing of the soil is occurring because either the contact pressure is too high or the moisture content in the soil is too high.

Pneumatic-tired Rollers

The contact pressure of pneumatic rollers depends on the weight of the roller and tire inflation pressure. Since about 10% of the load is carried by the tire side-wall, contact area may be estimated by using the following equation:

$$\text{contact area} = \frac{0.9 \times \text{wheel load}}{\text{tire pressure}} \tag{7-1}$$

The variation of vertical pressure with depth is illustrated in Figure 7-9. Since increasing roller weight at a constant tire pressure does not appreciably increase the contact pressure, the usual result is an increase in depth of compaction but little change in surface density. However, excessive tire pressure may result in shearing of the surface so that surface density actually decreases.

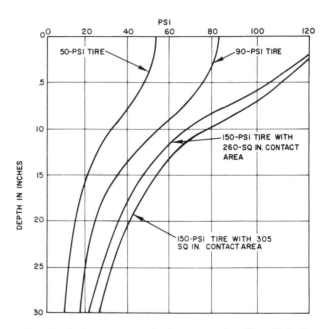

Figure 7-9 Vertical stresses vs. depth, pneumatic roller. (U.S. Department of the Army.)

Tests have again shown little relation between roller speed and compaction. Speed is usually limited by tractor power and job conditions.

Steel-wheeled Rollers

The contact pressure of steel-wheeled rollers is determined by the extent to which the soil deforms under the roller as well as by the weight of the roller. If roller weight is too great, soil displacement may be excessive. To minimize lateral displacement of the soil, rolling should proceed from low to high on sloped surfaces. As with other rollers, rolling speed has little effect on density. However, displacement in soft soils may limit operating speed.

Vibratory Compactors

The effectiveness of vibratory compactors depends on the frequency and amplitude of vibration as well as on their static weight. In general, vibratory compactors weigh only about one-half as much as comparable nonvibratory units. Vibration is most effective when its frequency approaches the resonant frequency of the soil being compacted. The density obtained with a vibratory compactor depends on the travel speed and the number of passes completed. However, for granular soils, density approaching maximum density may be obtained after only two to three passes. The allowable speed is determined by the soil type and desired depth of compaction as well as vibrator characteristics. Vibratory compactors are most effective on granular, noncohesive soils. However, at low frequency and high amplitude they are also effective in cohesive soils.

Job Management

After appropriate compaction equipment and operating methods have been selected (using the preceding guidelines and field tests), the manager's principal job is to insure that the operating plan is followed. The measure of success is, of course, the extent to which design specifications are met. A rapid method of checking densities during operation is provided by the nuclear density device mentioned earlier.

An additional factor to be considered during job planning and equipment operations is traffic control. Hauling equipment must be given the right-of-way. The use of high-speed compaction equipment may be required in order to keep up with the hauling equipment so that traffic bottlenecks are not created.

It should be apparent that the actual production achieved in a compacted fill is limited to the lower value of the hauling fleet capacity and the compaction fleet capacity.

7-4 PRODUCTION

Production Estimation

Compactor production in cubic yards per hour may be estimated by using Equation 7-2. However, accuracy of the results depends on the accuracy with which the rolling speed and number of passes required to obtain compaction can be estimated. In many cases, trial operations under similar soil and job conditions will be required to obtain accurate estimates. Table 7-4 lists typical operating speeds for common compactors.

$$\text{production (CCY/hr)} = \frac{16.3 \times W \times S \times L \times E}{P} \qquad (7\text{-}2)$$

where

$P =$ number of passes required;
$W =$ width compacted per pass (ft);
$S =$ compactor speed (mph);
$L =$ compacted lift thickness (in);
$E =$ job efficiency.

TABLE 7-4 *Typical operating speed of compaction equipment*

Computer	Speed (mph)
Sheepsfoot or tamping foot, crawler towed	3–5
Sheepsfoot or tamping foot, wheel tractor towed	5–10
High-speed tamping foot:	
First 2–3 passes	3–5
Walking out	8–12
Final passes	10–14
Heavy pneumatic	3–5
Multi-tired pneumatic	5–15
Grid roller:	
Crawler towed	3–5
Wheel tractor towed	10–12
Smooth wheel	2–4

Power Required

The power required for towing compaction equipment depends on the equipment's total resistance (grade plus rolling resistance). For compacting 6-in. thick layers (lifts) of average soil, the rolling resistance of sheepsfoot rollers has been found to be about 500 lb/ton of roller weight. Tamping foot rollers have a slightly lower rolling resistance. The rolling resistance of pneumatic-tired rollers may be found by using the methods presented in Chapter 5.

PROBLEMS

1. Calculate the rolling resistance of a 50-ton pneumatic roller if the tires penetrate 2 in. into the surface.

2. Estimate the rolling resistance of a 6-ton sheepsfoot roller in average soil.

3. Estimate the production in compacted cubic yards per hour for a tamping foot roller under the following conditions: average speed = 10 mph, compacted lift thickness = 6 in., roller width = 10 ft, job efficiency = 80%, number of passes = 12.

4. The data below were obtained from a standard AASHO compaction test. What is the optimum moisture content and 100% AASHO standard density of this soil?

Dry Density (pcf)	Moisture Content (%)
100	12
102	14
105	16
105	18
102	20
100	22

5. Select the appropriate roller and lift thickness for each of the following soils:

   Heavy clay (CH)

   Silty clay (CL)

   Well-graded gravel (GW)

   Clayey sand (SC)

6. A highway embankment was being constructed from a silty clay soil. The specifications required that the soil be compacted to 95% of standard AASHO density. However, compaction with a pneumatic roller with a tire pressure of 60 psi was producing only 92% to 94% of the standard AASHO density, and fill appeared to be a little spongy in places. The specifications permitted the moisture content of the fill to vary within $\pm 2\%$ of the laboratory optimum and the field moisture content was within these limits. The contractor brought in a heavier pneumatic roller with a tire pressure of 100 psi in an attempt to correct the situation. Although the use of the heavier roller produced the required density, i.e., 95% of laboratory maximum, the fill became completely unstable, weaving and rutting under the action of the roller and other compaction equipment. Analyze this situation, giving causes of the difficulty, and suggest corrective measures. Comment on the specifications for controlling water content.

7. A low plasticity clay is being placed in a highway embankment. The natural moisture content of the material is slightly below the standard AASHO optimum. The specifications require that some type of tamping foot roller be used for compaction. However, the roller being used does not always "walk out," and even when it does, it does not always produce the required dry density. An argument has developed between the project engineer and the contractor over corrective measures. The contractor wishes to add more ballast to the roller, thus increasing the contact pressure of the feet. The project engineer suggests that the weight of the roller is adequate and that the foot size should be increased. Which recommendation is more nearly correct? Why?

8. A highway contractor opened a cut in a very fine silty sand material. The cut was operated in such a way that surface water drained into the excavation and the cut material was very wet. Excavation and hauling were accomplished with very large scraper units aided in excavation by a crawler type dozer operated as a pusher.

Compaction of this material in an adjacent fill was with a heavy sheepsfoot roller pulled by a tracked bulldozer and operating on loose lift thicknesses of 8–10 in. It was apparent that compaction (specified to be 95% standard AASHO maximum dry density) was not being attained. The sheepsfoot roller did not "walk out," and the tires of the scrapers caused deep rutting and massive displacements of the embankment material as they traveled over it.

What suggestion would you make to the contractor for improving the compaction of the embankment? Explain.

REFERENCES

1. "Break out of the Compaction Squeeze," *Construction Equipment*, Vol. 24, No. 2 (February 1973), pp. 21–31.

2. BROMS, B., and L. FORSSBLAD, *Classification of Soils with Reference to Compaction.* Stockholm, Sweden, Swedish Geotechnical Institute, 1968.

3. BROWN, V., *Soil Compaction in Narrow Places.* Denver: Arrow Manufacturing Co., 1967.

4. *Compaction Handbook.* Portland, Oregon, Hyster Co., 1966.

5. FORSSBLAD, L., "Investigations of Soil Compaction by Vibration," *Acta Polytechnica Scandinavia*, No. Ci 34, 1965.

6. *Fundamentals of Compaction.* Peoria, Illinois: Caterpillar Tractor Co., 1971.

7. *Handbook of Compactionology.* Minneapolis: American Hoist and Derrick Co., 1968.

8. JOHNSON, A. W. and J. R. SALLBERG, "Factors Influencing Compaction Test Results," *Highway Research Board Bulletin 319.* Highway Research Board, Washington, D.C., 1962.

9. *TM5-331A: Earthmoving, Compaction, Grading and Ditching Equipment.* Washington, D.C.: U.S. Department of the Army, 1967.

10. TOWNSEND, D. L., *The Performance and Efficiency of Standard Compacting Equipment*, Engineering Report No. 6. Kingston, Ontario: Queens University, 1959.

8

GRADERS AND FINISHING

8-1 GRADING AND FINISHING

General

Grading is bringing earthwork to the desired shape and elevation. Finishing is a term used in roadwork for bringing earthwork to the final shape and grade required by plans, smoothing slopes, and shaping ditches. Finishing normally follows closely behind excavation and compaction. The motor grader is the most widely used machine for grading and finishing, although the use of specialized finishing machines called *grade excavators* or *grade trimmers* has increased in recent years.

The quantity of earthwork involved in finishing is not usually measurable, and finishing is seldom a pay item in a contract. Thus, there has been a tendency for planners and supervisors to provide only a cursory analysis of finishing and concentrate on excavation and fill (which are measurable pay items). However, studies have shown that the care with which finishing operations are planned and executed may greatly increase or decrease profits on an earthwork job.

Finishing in Road Work

Although highway pavement designs vary widely depending on materials used, design loads, and climatic conditions, typical roadway components are illus-

144

trated in Figure 8-1. Each layer of the roadway structure must be shaped to the required levels before the succeeding layer is placed.

The process of cutting down high spots and filling in low spots is called *balancing*. The final shaping of each roadway layer to specifications is called *trimming*.

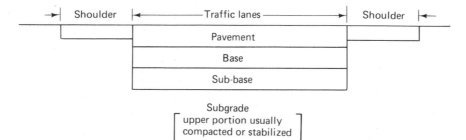

Figure 8-1 Typical roadway components.

Typical final grade tolerances allowed in highway work are:

Subgrade: $\frac{1}{2}$ to 1 in./10 ft

Subbase: $\frac{1}{4}$ to $\frac{1}{2}$ in./10 ft

Base: $\frac{1}{8}$ in./10 ft

Tolerances are even lower for some airport runways and high-level highways.

Economics of Haul Road Maintenance

The necessity for maintaining haul roads in good condition was pointed out in Chapters 4 and 6. Caterpillar Tractor Company has run cost studies comparing the performance of haul roads without maintenance with those maintained by motor graders. One such study involved a 2,000-ft haul road with varying grades and rolling resistance. The soil was a clay material weighing approximately 3,000 lb/BCY. Four different types of scrapers were used with and without motor grader support. The results of the study indicated that maintenance of the haul road by motor graders yielded savings of 0.8¢ to 1.3¢ per bank cubic yard of excavation as well as a production increase of at least 20%. Although it would be unusual to operate a haul road with absolutely no maintenance, these results do point up the increase in efficiency that may be achieved by providing adequate grader support.

8-2 GRADERS

Characteristics

The modern motor grader is a multi-purpose machine used for grading, shaping, bank sloping, ditching, and scarifying as well as the mixing and spreading of soil and bituminous materials. It is used in general construction and in the construction and maintenance of highways. It may also be used for light stripping operations and for snow removal.

A typical motor grader is illustrated in Figure 8-2. The blade is referred to as the *moldboard* and is equipped with a replaceable cutting edge and end bits.

Figure 8-2   Motor grader. (Courtesy WABCO Contruction and Mining Equipment Group.)

The grader moldboard may be placed in a wide range of positions for different types of operations as illustrated in Figure 8-3. The moldboard pitch may also be varied as shown to obtain either a cutting or a dragging action. Forward pitch results in a dragging action which is used for light cuts and for blending material. Backward pitch increases cutting action but may allow material to spill over the blade.

The front wheels of a motor grader may be leaned to either side to offset the side thrust produced by soil pressure against the angled blade and to assist in turning the grader. Articulated frames are also used on some graders to increase grader maneuverability and versatility. The three modes of operation possible with an articulated grader are illustrated in Figure 8-4. In the straight mode (A) the machine

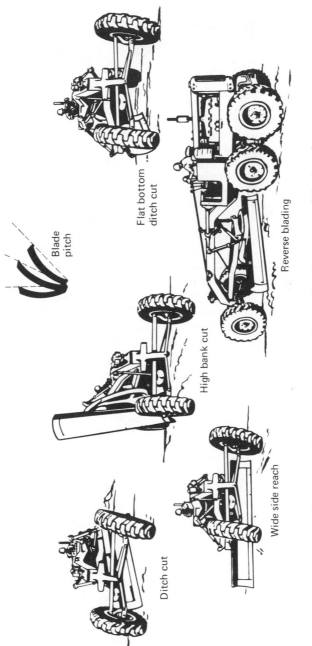

Flat bottom
ditch cut

Blade
pitch

High bank cut

Reverse blading

Ditch cut

Wide side reach

Figure 8-3 Motor grader blade positions. (U.S. Department of the Army.)

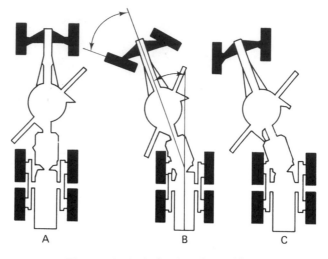

Figure 8-4 Articulated grader positions.

operates in the conventional manner. In the articulated mode (B) the machine can achieve a very short turning radius. In the crab mode (C) the front and rear wheels are offset so that the driving wheels can remain on firm ground while the machine is cutting banks, side slopes, or ditches.

Also available are graders with automatic blade control systems that allow precise grade control. A sensing system is used that follows an established surface or string line and that automatically raises or lowers the moldboard to achieve the desired grade. This grader, illustrated in Figure 8-5, is often referred to as an *ABC grader*.

Grader Operations

Spreading

Graders may be used for spreading materials. However, the load which can be carried on the blade is limited by the power and traction of the grader as well as by the height of the moldboard. Thus, the amount of material which can be pushed is considerably less than that which a dozer can push. However, the blade capacity is increased somewhat by the concave shape of the moldboard which imparts a rolling action to the material. Material to be spread by the grader should be pre-spread as much as possible during dumping in order to reduce the height of material piles.

Sidecasting

Material may be moved laterally (or sidecast) by a grader when the moldboard is set at an angle which is not perpendicular to the direction of travel. In this position the material drifts off the rear end of the blade to form a longitudinal pile or windrow.

Figure 8-5 Grader with automatic blade control. (Courtesy CMI Corp.)

Windrows should not be allowed to form in line with the rear wheels because the material would reduce traction as well as change the moldboard cutting position. For usual road shaping and maintenance, the moldboard is set at an angle of 25° to 30° from the perpendicular position. For spreading, this angle would be decreased. Hard cutting and ditching require an increased blade angle. Sidecasting may also be used to fill ditches or cover pipelines after installation.

Planing

Planing or fine grading is accomplished by setting the moldboard at a slight angle to the perpendicular and at an elevation that cuts off peaks and allows the depressions to be filled. To accomplish this, a partial load of material must be kept in front of the moldboard at all times.

Ditching

Graders make both V-type and trapezoidal ditches. The maximum economical ditch depth is about 3 ft with a maximum bottom width of about 4 ft. Ditches

exceeding these dimensions are better constructed by using scrapers or one of the
excavators described in Chapters 2 and 3. Ditching during road construction is
one of the basic uses for the grader. A suggested procedure for this operation is as
follows:

> Make a 3 to 4 in. deep marking cut at the outer edge of the bank slope. During
> this cut the forward end of the blade should be in line with the outside edge of the
> front tire as it travels the ditch line. Follow the marking cut with the required
> number of ditching cuts. Cuts should be as deep as can be made accurately and
> without overloading the grader. Succeeding cuts should be moved in toward the
> road so that the desired angle is obtained on the outer ditch slope and the final
> cut falls along the desired ditch bottom line. Material from ditching cuts forms
> a windrow under the grader and is then sidecast onto the road or hauled away.
> After the ditching cuts are complete, the bank slope is shaped to its final grade.
> The ditch is then cleaned with excess material being cast onto the shoulder. Upon
> completion of ditching, final shaping of the shoulder and roadway are performed.

Scarifying

The grader scarifier is used to increase blade efficiency in tough material and
for breaking up asphalt pavements. For operation in hard material, it may be necessary to remove some scarifier teeth to reduce the load on the machine. This is similar
to the procedure used in ripping which will be covered in Chapter 9. However,
scarifiers are not designed for heavy-duty jobs and should not be used as substi-
tutes for rippers. Rippers are also available for graders and are usually mounted on
the rear end of the grader. Rippers can cover a wider path and rip to a greater
depth than scarifiers. The ripping ability of a grader is limited, however, by the
grader's weight and power.

Long shanks are available for scarifiers. These should be used to increase
aeration when drying out moist materials or when material packs up in front of the
scarifier during the scarifying of asphalt pavements and other cohesive materials.

Moldboard Edges

Flat or curved cutting edges are available in varying heights and thicknesses
for moldboards. Serrated edges are also available for work in frozen soil, ice, and
packed gravel. Curved edges are recommended for fine grading and cutting hard
materials. Flat edges are recommended whenever edge wear is important and penetration is not difficult. Thin blades give better cutting action, but thicker blades
give more wear life.

Extra-wide (high) cutting edges are available for maximum life under severe
wear conditions.

In addition to replaceable cutting edges, replaceable end bits are used on the
moldboard to prevent damage to the ends of the moldboard. Also available are
overlay bits that fit behind the junction of the cutting edge and end bit to strengthen
moldboard corners and limit corner wear.

Estimating Grader Production

Grader production is usually estimated on an area basis (square feet or square yards per hour) or linear basis (miles completed per hour).

The time required to complete a job may be calculated on a linear basis by using the following equation:

$$\text{time (hr)} = \frac{\text{number of passes} \times \text{length (mile)}}{\text{average speed (mph)} \times \text{efficiency}} \qquad (8\text{-}1)$$

Average speed obtained is a function of the type of work, machine characteristics, and operator skill. Table 8-1 lists approximate grader speeds on various tasks for a typical grader.

TABLE 8-1 *Typical grader speeds*

Type of Work	*Speed (mph)*
Ditching	2.5–4.0
Bank sloping	2.5
Road construction and maintenance	4.0–6.0
Finishing	4.0–9.0
Spreading	6.0–9.0
Mixing	9.0–20.0
Snow removal	12.0–20.0

Example 8-1

PROBLEM: Ten miles of gravel road require reshaping and leveling. It is estimated that six passes of a grader will be required. Based on material, machine, and operator skill you estimate two passes at 3.0 mph, two passes at 4.0 mph, and two passes at 5.5 mph will be required. Job efficiency is estimated at 0.75. How many grader hours will be required for this job?

Solution:

$$\text{Average speed} = \frac{(2 \times 3.0) + (2 \times 4.0) + (2 \times 5.5)}{6} = 4.17 \text{ mph.}$$

$$\text{Time required} = \frac{6 \times 10}{4.17 \times 0.75} = 19.2 \text{ hr (Equation 8-1).}$$

Job Management

The following suggestions are made for obtaining maximum grader production efficiency:

1. Use only the minimum number of passes required to perform the work. Careful job planning, supervision, and skilled operators are required here.

2. Eliminate as many turns as possible. Usually, it is more efficient for a grader to back up rather than turn around for passes of less than 1,000 ft. Grading

in reverse may be used for long passes if the operator is skilled and if turning is difficult or impossible.

3. Graders may be worked in tandem if space permits and if machines are available. This technique is especially useful in leveling, spreading, and mixing operations.

4. Table 8-2 provides a checklist for insuring efficient performance in various grader operations.

8-3 GRADE EXCAVATORS AND TRIMMERS

The grade excavator or grade trimmer can finish subgrades and bases with greater accuracy and at a higher speed than can motor graders. Excavators and trimmers are often capable of scarifying soil and tearing up old asphalt pavements as well as excavating and trimming. Most models are equipped with integral belt conveyors for loading excess material into haul units or placing it in windrows outside the excavation area. A large grade trimmer/reclaimer which can also compact base material, lay asphalt, and act as a slipform paver is shown in Figure 8-6.

Figure 8-6 Large grade trimmer/reclaimer/paver. (Courtesy CMI Corp.)

TABLE 8-2 *Checklist for efficient grader operations*

Leveling	Ditching	Reshaping	Bank Sloping	Scarifying	Mixing	Spreading
1. Is blade set at sharp enough angle to get proper slicing action?	1. Is marking cut made approximately 3 to 4 in. deep?	1. Is maintenance performed in least possible number of passes?	1. Is heel of blade at bottom of ditch?	1. Are teeth penetrating too deep so subgrade is torn up and mixed with road surface?	1. Is blade pitched forward to obtain rolling action?	1. Are there any signs that material is not thoroughly mixed?
2. Is blade set at proper pitch to get cutting action?	2. Is toe of blade directly behind front wheels?	2. Are shoulders and edges maintained to eliminate standing water?	2. Is blade set to give maximum slicing action?	2. If material is hard and consolidated, are enough teeth removed to give proper penetration?	2. Are graders worked in tandem where possible?	2. Is material being spread to proper thickness?
3. Is blade set so windrow falls either inside or outside of rear wheels?	3. Is blade set so windrow falls inside of rear wheels?	3. Is material kept on road and not spilled into ditch?	3. Is bank sloping being done in first gear?	3. Is machine traveling as fast as power allows?	3. Is blade taking too much material, preventing free rolling?	3. Are wheels being turned on freshly spread mix, causing a wavy surface?
	4. Is windrow removed before it reaches underside of grader?	4. Is speed too fast or cut too deep, causing washboarding of surface?	4. Does the rear wheel trail heel of blade?		4. Is speed regulated to give proper mixing action and at the same time to allow full control of grader?	
	5. Is bank slope step-cut?	5. Is speed fast enough to maintain efficient operation?			5. Is blade set to produce a free flow of material?	
	6. Is the heel of the blade lowered to maintain inside slope of ditch?	6. Has operation been planned so flow of traffic is not stopped?				

(U.S. Department of the Army[5])

Because of their size and low travel speed these machines are usually transported between jobs on heavy equipment trailers. They often also require partial disassembly to meet highway size limitations during transportation. Although these machines lack the versatility of motor graders, they are very effective when used properly. In addition to yielding higher production, they may be more economical than graders in a given situation. Their principal advantages, however, are high production rates and precise grade control.

PROBLEMS

1. Explain the meaning of the following terms used in connection with motor graders:
    (a)   Moldboard.
    (b)   Blade pitch.
    (c)   Articulated grader.
    (d)   Scarifier.
    (e)   End bit.

2. Estimate the production in linear feet per hour that a motor grader will achieve while performing bank sloping under the following conditions:
    (a)   Average number of passes required = 6.
    (b)   Job efficiency = 0.60 (including turning time).

3. Which blade angle and pitch should be used for mixing material?

4. Sketch a cross section of a V-type ditch showing the material removed by each succeeding cut of a grader.

5. Thirty miles of gravel road require reshaping and leveling. You estimate that two passes at 3.5 mph, two passes at 4.5 mph, and one pass at 6.0 mph will be required for a motor grader. How many hours of grader operation should this job require if the job efficiency is 0.80?

REFERENCES

1. *Basics of Finishing*. Peoria, Illinois: Caterpillar Tractor Co., n.d.

2. *Engineering Bulletin RB-224A: Road Building Equipment*. Chicago: American Oil Company, 1962.

3. HAVERS, JOHN A., and FRANK W. STUBBS, JR., eds., *Handbook of Heavy Construction*. New York: McGraw-Hill, 1971.

4. *Motor Grader Economics*. Peoria, Illinois: Caterpillar Tractor Co., n.d.

5. *TM5-331A: Earthmoving, Compaction, Grading and Ditching Equipment*. Washington, D.C.: U.S. Department of the Army, 1967.

9

ROCK EXCAVATION

9-1 TRADITIONAL METHODS

Introduction

Traditionally, rock excavation has been accomplished by drilling blastholes, loading them with explosives, detonating the explosive, and then loading and hauling away the fractured rock. In recent years the development of heavy-duty tractors, rippers, and hauling equipment has provided a new and often more economical alternative which will be discussed later in the chapter. First, let's briefly review the equipment and procedures used in the traditional methods of rock excavation.

Drilling

Most blasthole drilling is accomplished by percussion-type drills powered by compressed air. Drills range from the hand-held rock drill (or jackhammer) through the wheel-mounted wagon drill to the crawler drill. Typical capacities for these drills are given in Table 9-1; representative drilling rates are shown in Table 9-2. Rotary-percussion drills which combine rotary and percussion drilling action are able to drill approximately three times as fast as similar percussion drills.

TABLE 9-1 *Typical percussion drill capabilities*

Type Drill	Maximum Drill Diameter (in.)	Maximum Depths (ft)
Hand drill	$2\frac{1}{2}$	10
Wagon drill	4	30
Crawler drill	6	100

TABLE 9-2 *Representative drilling rates*

Bit Size (in.)	Rock Class	Hand Drill Case Hardened Bit	Hand Drill Tungsten Carbide Bit	Wagon Drill Case Hardened Bit	Wagon Drill Tungsten Carbide Bit	Crawler Drill Case Hardened Bit	Crawler Drill Tungsten Carbide Bit
$1\frac{5}{8}$	Soft	19	24				
	Medium	13	20				
	Hard	8	14				
$1\frac{3}{4}$	Soft	15	21				
	Medium	10	16				
	Hard	5	11				
2	Soft	13	19	30	50		
	Medium	8	14	25	30		
	Hard	4	10	15	25		
$2\frac{1}{8}$	Soft			28	50		
	Medium			23	30		
	Hard			13	22		
$2\frac{1}{4}$	Soft			27	48		
	Medium			22	28		
	Hard			12	20		
$2\frac{1}{2}$	Soft			27	48		
	Medium			22	28		
	Hard			11	19		
$2\frac{3}{4}$	Soft			26	46		181
	Medium			21	26		143
	Hard			10	18		90
3	Soft			26	46		159
	Medium			21	26		114
	Hard			10	18		75
4	Soft			15	35		123
	Medium			10	20		87
	Hard			2	10		50

(U.S. Department of the Army[5])

There are also several types of pure rotary drills available. Of these, only the blasthole drill (which uses a roller bit) is typically used in blasting work.

Fusion piercing is a relatively new technique using a jet torch combined with a water spray to penetrate rock. Rock heated by the jet is cooled suddenly by a water spray, causing fracturing of the rock.

The size, depth, and spacing of blastholes and the amount and type of explosive used are determined by rock conditions and the type of fracturing desired. Although equations for estimating the required explosive charge and spacing are available, test blasts are usually required to determine optimum hole and explosive characteristics.

Blasting

After the blastholes have been drilled, they are loaded with the required amount of explosive, detonators are attached, and holes are stemmed. Stemming is the process of sealing the hole with inert material to increase the effectiveness of the explosive.

Although dynamite is sometimes still used for rock excavation, recent developments in the use of ammonium nitrate and fuel-oil compounds have reduced the cost of explosives for rock breakage as much as 75% below the cost of dynamite.

Volume of Rock Removed

The volume of rock removed by one blast may be measured by cross sectioning the material before and after excavation or by weighing the rock hauled away. When a vertical rock face is present (as in a quarry), the volume of rock removed by a blast may be estimated by multiplying the blasthole pattern length by its width by the depth of hole.

Excavation Cost

The cost per cubic yard of rock excavated is, of course, the total cost of blasting, loading, and hauling divided by the volume of rock removed. Total cost must include all elements of the operation. These costs must include the following:

1. Equipment cost, including drills, compressors, and auxiliary equipment, as well as loaders and haulers

2. Drill bit cost

3. Explosive cost, including storage and transportation

4. Material used for stemming

5. Labor costs, including any layout work

6. Cost of liability and property damage insurance

Tunneling

Tunneling is a highly specialized form of rock excavation that will not be covered in detail here. The procedure, however, has traditionally been similar to the drilling and blasting operations previously discussed. The basic sequence is drill, blast, and muck out (remove) the blasted rock. Drilling has traditionally used percussion drills powered by compressed air. The drill used is frequently a heavier model of the hand drill, called a *drifter*. Drifters are often mounted on a large frame called a *jumbo* which enables the drillers to work on the entire face of the tunnel at one time. After being blasted, the loose rock is loaded by shovels, special mucking shovels, or other loading equipment onto rail cars, trucks, or conveyor belts for movement out of the tunnel.

The use of large mechanical tunneling devices called *moles* has increased rapidly in recent years. Other new and promising methods for speeding up tunneling operations include using a plasma torch and the firing of inert projectiles into the rock.

9-2 ROCK CHARACTERISTICS

Types of Rock

The degree of difficulty involved in fracturing and excavating rock depends on a number of factors. A major factor is the type of rock involved. There are three basic classifications of rock: igneous, sedimentary, and metamorphic. Igneous rock was formed by the cooling of the earth's molten material. Because of its method of formation it is rather homogeneous and has few planes of weakness. Sedimentary rock was formed by the precipitation of material from water or air. Hence, it is highly stratified and can usually be fractured rather easily along its planes of stratification. Metamorphic rock is igneous or sedimentary rock which has been changed in composition or texture as a result of physical or chemical forces. Thus, the difficulty of excavation is usually intermediate between that of igneous and sedimentary rock.

Characteristics Affecting Excavation

In addition to the basic type of rock involved, other characteristics affecting the difficulty of excavation include the following:

1. The extent of stratification, fractures, and planes of weakness

2. The amount of weathering

3. The brittleness of the rock

4. The predominant grain size and whether or not the material has a crystalline nature

Rock Investigations

Although many of the above characteristics are determined by visual observation, only surface conditions are observable without further investigation. Subsurface conditions are determined by removing samples with a core drill, by digging a test pit large enough to permit visual observations (by persons or by TV cameras), or by making seismic measurements.

The use of the refraction seismograph to determine rock conditions at shallow depths has been well developed in recent years. The method of analysis is based on the velocity of sound traveling through the soil and rock. Sound velocity is measured by the time required for a geophone receiver to pick up the sound waves created by a small explosion or a hammer at a known distance from the receiver. This velocity has been found to range from about 20,000 fps in sound hard rock down to about 1,000 fps in a loose soil.

It has also been found that the refraction seismograph method permits determination of the thickness of the rock and soil layers as well as their hardness. To determine layer thickness, sound sources or geophones are located at fixed intervals from each other. The measured time of travel is then plotted against distance between the sound source and the receiver. When layers are present, the results will plot as straight lines with different slopes for each layer. The greater the slope of the line, the softer the material. However, it should be noted that the equations given below for computing layer thickness are valid only when the sound velocity increases with layer depth, i.e., V_1 is greater than V_1, V_3 is greater than V_2, and so on.

The thickness of the upper layer of a soil/rock system may be computed by using Equation 9-1.

$$H_1 = \frac{D_1}{2} \sqrt{\frac{V_2 - V_1}{V_2 + V_1}} \tag{9-1}$$

where

H_1 = thickness of upper layer (ft);
D_1 = distance from sound source to first intersection of lines on time–distance graph (ft);
V_1 = velocity in upper layer (fps);
V_2 = velocity in second layer (fps).

In most construction applications we are primarily concerned with the thickness and hardness of the upper soil/rock layer and with the hardness of the second layer. However, in some cases, it may also be desirable to determine the thickness of the second layer and the hardness of the underlying (third) layer. This can be accomplished by using Equation 9-2.

$$H_2 = \frac{D_2}{2} \sqrt{\frac{V_3 - V_2}{V_3 + V_2}} + H_1 \frac{V_3 \sqrt{V_2^2 - V_1^2} - V_2 \sqrt{V_3^2 - V_1^2}}{V_1 \sqrt{V_3^2 - V_2^2}} \tag{9-2}$$

where

H_1 = thickness of upper layer (ft)
H_2 = thickness of second layer (ft);
D_2 = distance from sound source to second intersection of lines on time–distance graph (ft);

V_1 = velocity in upper layer (fps);
V_2 = velocity in second layer (fps);
V_3 = velocity in third layer (fps).

Equation 9-3 provides a simpler method for calculating the depth of the second layer. Although this equation yields only an approximate value, its accuracy is comparable to the accuracy of the field measurements obtained by using a portable seismograph.

$$H_2 = \frac{D_2}{2}\sqrt{\frac{V_3 - V_2}{V_3 + V_2}} - 0.2H_1 \qquad (9\text{-}3)$$

The use of Equations 9-1 and 9-2 is illustrated in Example 9-1.

Example 9-1

PROBLEM: Find the seismic wave velocity and depth of the upper two rock layers based on the following refraction seismograph data:

Distance from Sound Source to Geophone (ft)

	10	20	30	40	50	60	70	80	90	100
Time (msec)	5	10	15	20	23	27	30	32	34	36

Solution:

Plot time of travel against distance from sound source to geophone as shown below.

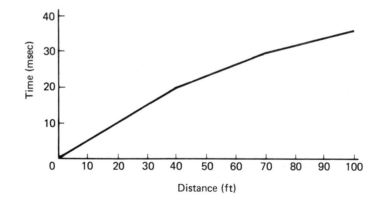

$V_1 = \dfrac{40}{.020} = 2{,}000$ fps.

$V_2 = \dfrac{70 - 40}{.030 - .020} = 3{,}000$ fps.

$V_3 = \dfrac{100 - 70}{.036 - .030} = 5{,}000$ fps.

$$H_1 = \frac{40}{2}\sqrt{\frac{3,000 - 2,000}{3,000 + 2,000}} = (20)(0.2)^{1/2} = 9.0 \text{ ft (Equation 9-1)}.$$

$$H_2 = \frac{70}{2}\sqrt{\frac{5,000 - 3,000}{5,000 + 3,000}}$$
$$+ (9)\frac{5,000\sqrt{3,000^2 - 2,000^2} - 3,000\sqrt{5,000^2 - 2,000^2}}{2,000\sqrt{5,000^2 - 3,000^2}}$$

(Equation 9-2).

$$H_2 = (35)(0.5) + (9)\frac{(5,000)(2,236) - (3,000)(4,583)}{(2,000)(4,000)}.$$

$$H_2 = 17.5 - 2.9 = 14.6 \text{ ft}.$$

9-3 ROCK RIPPING AND HAULING

Rock Ripability

Because of the availability of heavy-duty tractor mounted rippers in recent years it has often become feasible to rip rock for excavation instead of resorting to drilling and blasting. The best method of determining ripability is by running tests in the field. However, observation on a large number of jobs has yielded a good correlation between seismic velocity of rock and its ripability. Figure 9-1 indicates ripper performance which might be expected for a particular tractor-ripper combination in various types of rock as a function of seismic velocity.

Ripping Equipment

Rippers have been used for centuries to break up hard soils. The Romans, for instance, used a form of ripper towed by oxen. Rippers towed by tractors are still used today. A towed ripper is also called a *rooter*. The tractor mounted ripper, however, has largely displaced the towed ripper. The tractor mounted ripper is able to apply much more downward pressure to the ripper than can the towed ripper. In addition, the tractor mounted ripper can be permanently attached to the tractor without interfering with the tractor's use as a dozer. A tractor mounted ripper is shown in Figure 9-2.

There are three basic types of tractor mounted ripper, as illustrated in Figure 9-3. The hinge-type or radial lift ripper (A) pivots about its point of attachment so that the tooth angle in relation to the soil surface changes with ripper penetration. The parallelogram-type or parallel lift ripper (B) maintains a constant tooth angle as it is raised and lowered. The adjustable parallelogram-type ripper (C) allows the tooth angle to be varied as desired to obtain optimum results, even while moving. While the optimum choice of ripper will vary with rock and job conditions,

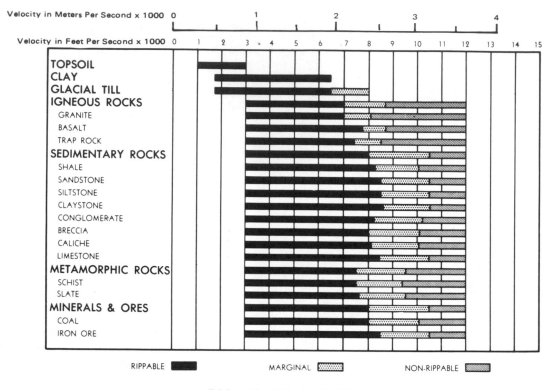

D9G — No. 9 Series D RIPPER

Figure 9-1   Ripper performance vs. seismic velocity. (Courtesy Caterpillar Tractor Co.)

the adjustable parallelogram ripper is the most versatile and can be used in any application.

Ripper shanks come in a variety of designs, each best suited for certain applications. Replaceable ripper tips and shank protectors are commonly used. Ripper tips are also available in different shapes and lengths to suit the material involved.

Tandem Ripping

When ripping conditions are marginal, the use of tandem tractors will often result in a substantial increase in production. In such a case, the unit cost of production would be reduced by tandem ripping. The use of a dual tractor described in Chapter 5 would be even more effective than the use of tandem tractors. An approximate relation between tandem ripping and single-tractor ripper production is illustrated in Figure 9-4.

Figure 9-2 Tractor with heavy duty ripper. (Courtesy Caterpillar Tractor Co.)

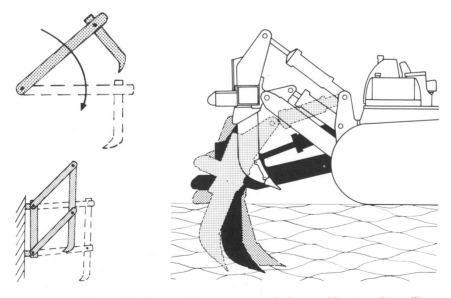

Figure 9-3 Types of tractor mounted rippers. (Courtesy Caterpillar Tractor Co.)

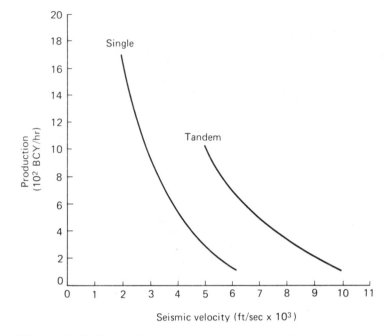

Figure 9-4 Typical tandem vs. single tractor ripper performance.

Estimating Ripper Production

Methods of measuring production during rock excavation were discussed in Section 9.1. An estimate of ripper production can be made by using the following equation:

$$\text{production (BCY/hr)} = \frac{2.22 \times D \times W \times L \times E}{T} \qquad (9\text{-}4)$$

where
  D = average penetration (ft);
  W = average width loosened (ft);
  L = length of pass (ft);
  E = job efficiency factor;
  T = ripper cycle time, including turns (min).

Loading and Hauling Ripped Rock

Along with the development and use of tractor mounted rippers for fracturing rock have come new developments in rock loading and hauling equipment. Rubber-tired front-end loaders equipped with heavy-duty rock buckets and extra deep tread tires (or chains) often prove to be more economical than the traditional power shovel. Other advantages of the wheel loader over the shovel include increased mobility on the job site, ability to operate in a limited space, versatility, and ease of movement between jobs. Another loading method sometimes used involves dozers to push the rock into hoppers or onto belts for loading trucks or wagons. Scrapers are also used for both loading and hauling rock.

The most commonly used piece of equipment for hauling rock is the dump truck. Current trends in rock hauling equipment include the use of larger and faster trucks with rock bodies and the use of rock wagons. The major hauling development in recent years, however, has been the increased use of special model scrapers for operating in rock.

These scrapers have been especially reinforced for use in rock loading and hauling. Studies have shown that scrapers may be more economical than the shovel/truck combination for rock loading and hauling. In general, shorter haul distances favor the scraper while long haul distances favor the shovel/truck combination. In addition to possibly lower unit cost of production, scrapers often require a lower equipment investment than a comparable shovel/truck fleet. Scrapers may also accomplish better spreading and compacting of material in the fill than would dump trucks.

9-4 RIPPING OPERATIONS

Job Management

Some of the factors involved in ripping operations and some suggested job techniques are discussed below. As in the case of other rock excavation procedures, the optimum procedure is usually developed from trial operations under the particular job conditions. It should be remembered that the objective in rock ripping is to *shatter* the rock sufficiently for excavating and loading.

Operating Speed

In ripping rock, best results are usually obtained by operating in first gear. Tooth wear increases rapidly with increasing speed. In soft materials the higher speed of wheel dozers or graders may be utilized advantageously in ripping.

Number of Shanks

In general, use the number of shanks that obtain the desired penetration without undue strain on the tractor. When two shanks are used, the tractor tends to swerve to one side when one shank encounters a hard spot. This action is minimized by using a ripper with the two shanks mounted rather close together (sometimes called a *narrow gauge ripper*). Swiveling shank holders also help to reduce the load on the tractor when it turns or hits hard spots.

Depth and Spacing

The depth of shank penetration is largely determined by the available power and the number of shanks used. When working in stratified material penetration should match the depth of layers when possible. The lateral spacing of ripping passes depends on the hardness of material and the desired size of fractured material. In some materials, particularly thinly laminated shales, production is often increased by using a larger number of shanks at shallow depth instead of obtaining maximum penetration with a single shank.

Direction of Ripping

Whenever possible, rip downgrade to take advantage of negative grade resistance. In stratified material the direction of ripping may have to be chosen to allow the tip to get under the laminations. Cross ripping may be required to break up hard materials that will not fracture adequately when they are ripped in only one direction.

Ripping Surface

Leave several inches of ripped material on the surface before making another pass. This soil layer provides a cushion for the tractor and increases the coefficient of traction.

Preblasting

Preblasting is the process of setting off small blast charges to loosen rock and provide a fracture line before performing additional blasting. Preblasting before ripping permits ripping rock that would otherwise be unrippable.

Other Ripping Applications

As mentioned earlier, rippers are also available for graders and front-end loaders. These are frequently used for ripping old asphalt pavements or for breaking up hard soil to facilitate scraper and grader operations.

Also available are back rippers that mount on the back of dozer blades. These act as rippers to loosen soil while the dozer backs up. Their use will increase the productivity of dozers in all but very loose soil. When the dozer moves forward, the ripper teeth pivot to the rear and drag loosely on the soil surface.

During clearing operations rooters and rippers are used to cut tree roots and pull tree stumps. They can also be used to rip up old concrete pavements by catching the edge of the slab with the curved portion of the shank and then lifting the ripper to break the slab.

PROBLEMS

1. Using the following data, find the seismic wave velocity and depth of the upper rock layer:

Distance to source (ft)	10	20	30	40	50	60	70	80
Time (msec)	2.5	5.0	7.5	10.0	11.5	13.0	14.5	16.0

2. Given the geophone data below, calculate the seismic wave velocity in each soil layer. Find the depth of each soil layer except the bottom layer.

Distance (ft)	10	20	30	40	50	60	70	80	90
Time (msec)	10	20	30	35	40	45	46	47	48

3. You have measured a seismic velocity of 5,000 fps in limestone. Would you expect this rock to be rippable by a D9G tractor with ripper? (See Figure 9-1.) Would you recommend using a single or a tandem ripper on this job? Why?

4. A field test indicates that a ripper can attain a speed of 2 mph with a penetration of 2 ft while loosening a width of 6 ft. If the length per pass is 500 ft, turning time at the end of each pass is 0.3 min, and job efficiency is 80%, what do you estimate as the production in bank cubic yards per hour for these conditions?

5. Estimate the cost per bank cubic yard of production for both single and tandem rippers whose production curves are shown in Figure 9-4. The rock has a seismic velocity of 5,500 fps. Owning and operating cost for each tractor (including rippers) is $50/hour.

REFERENCES

1. *ATECO Ripping Handbook*. Oakland, California: American Tractor Equipment Corporation, 1970.

2. DOBRIN, MILTON B., *Introduction to Geophysical Prospecting* (2nd ed.). New York: McGraw-Hill, 1960.

3. *Handbook of Ripping*. Peoria, Illinois: Caterpillar Tractor Co., 1972.

4. HAVERS, JOHN A., and FRANK W. STUBBS, JR., eds., *Handbook of Heavy Construction*. New York: McGraw-Hill, 1971.

5. *TM 5-331C: Rock Crushers, Air Compressors and Pneumatic Tools*. Washington, D.C.: U. S. Department of the Army, 1968.

10

CONCRETE EQUIPMENT

10-1 CONCRETE CONSTRUCTION

General

Portland cement concrete is widely used as a construction material in almost all types of construction. Applications range from foundations for small structures to massive concrete structures such as dams. The components of a concrete construction operation will include most, if not all, of the following operations: site preparation, excavation, form construction, batching, mixing, placing, finishing, and curing. These operations and the equipment involved are discussed in the following sections. Concrete paving operations are discussed in Section 10.6.

Site Preparation, Excavation, and Forming

Site preparation involves all the work required prior to excavation in order to prepare the job site for the start of construction. Operations in this category include building or improving access roads to the job site, clearing and draining the construction site, and laying out the job. Equipment and methods used for these operations include those discussed in Chapters 2–8. Drainage may also require the use of dewatering equipment such as well points or pumps. Job-site layout establishes the exact location of the facility to be constructed and provides grade stakes or markers showing the extent of cut and fill required during the excavation phase. Site preparation may include the layout and construction of stockpiles and equipment for on-site concrete operations. These phases of the operation will be discussed later.

The excavation phase involves cutting and filling of the job site to within about 6 in. of final grade. Excavation to final grade for forming and placing the concrete usually involves the use of grading and finishing equipment covered in Chapter 8. The stability of slopes must always be considered in excavation operations and may require the use of shoring, rock ties, or soil stabilization techniques.

Forms are used to hold plastic concrete in the desired shape until sufficient strength has been attained for the concrete to support itself and any expected loads. For on-grade and below-grade structures, the soil itself is frequently used to provide part of the necessary forming. Erected forming usually consists of job-constructed or prefabricated forms of wood, metal, plastic, or a combination of these materials. Construction equipment utilized is primarily the lifting equipment described in Chapter 2.

10-2 BATCHING AND MATERIALS HANDLING

Batching

Batching is the process of accurately proportioning cement and aggregate for delivery to the mixer. Concrete ingredients must be accurately proportioned to obtain concrete of uniform quality. Specifications usually require accuracy within 1% for cement and water quantities, 2% for aggregates, and 3% for any admixtures used. Hence, batching equipment must be accurate within these ranges for any volume of material to be batched. Water is usually added to the dry materials in the mixer. However, in some central mix plants the introduction of water into the mixer is also controlled by the batcher. For small construction mixers, batching is normally accomplished by manually loading the required quantities of cement and aggregate directly into the skip or hopper of the mixer. Although the material may be proportioned by either weight or volume, the weight method is preferred because of its simplicity and accuracy. Cement is usually measured and handled by the sack (94 lb).

Central batching plants are commonly used with central mix plants or transit mixers. These frequently consist of separate aggregate and cement batching units. Aggregate batching units consist of several bins which weigh and release the required quantity of each aggregate size for each batch. Batched material is loaded directly into a truck or onto a conveyor for movement to the mixer. Cement batchers provide similar proportioning of cement for each batch. Bulk cement is loaded into the bin and kept in a loose state by passing a constant supply of compressed air through the bin.

In setting up a batch plant, care must be taken to provide adequate space for material stockpiles as well as for hauling and handling equipment. When discharging into trucks, establish a traffic pattern that separates batch trucks and material haul units for maximum efficiency.

Material Handling

Efficient material handling is of utmost importance in efficient batching operations. For central plant operation, aggregates are usually moved from stockpiles to batching plants by elevators, conveyors, or clamshells. Bulk cement is usually blown through ducts from trucks or railroad cars into the cement bin of the batching unit. In smaller portable plants, aggregates may be handled by portable conveyors, clamshells, front-end loaders, trucks, etc. Cement may be handled in bags and loaded into the cement batcher by an internal loading system.

The equipment chosen for material handling must be able to provide the required volume of material at the required elevation. Efficient plant layout and traffic patterns are a must for efficient plant operations.

10-3 MIXING EQUIPMENT

General

To produce quality concrete, it is essential that the concrete be thoroughly mixed so that all ingredients are evenly distributed. Samples taken from random portions of properly mixed concrete will possess approximately the same unit weight, shrinkage characteristics, and consistency (slump) as well as air, cement, and aggregate content.

The classification system used for concrete mixers assigns a rating symbol to a mixer. The symbol consists of a number and a letter. The number indicates the manufacturer's rating of the number of cubic feet of wet concrete per batch that the machine will mix satisfactorily. The letter identifies the type of mixer: S is a construction mixer; E is a paving mixer; and M is a mortar mixer. Thus, a 16S mixer is a 16-cu ft construction mixer; a 34E mixer is a 34-cu ft paving mixer. A 10% overload is usually allowed by the rating system. Mixers are available in several types; trailer-mounted, stationary, transit, and paving mixers are the most commonly used types.

Stationary Concrete Mixing

Stationary mixing takes place at a fixed location, either at the job site or at a central mix plant. Trailer-mounted or permanently installed mixers are used which range from 2 cu ft to over 12 cu yd in capacity. Mixing action used includes tilting or nontilting revolving drums and open-top paddle mixing. A trailer mounted 2-cu yd mixer with batching equipment is shown in Figure 10-1.

Figure 10-1 Trailer-mounted concrete mixing plant. (Courtesy Johnson Division, Koehring Co.)

Although construction and highway mixing specifications vary, a minimum mixing time of 1 min for mixers of 1 cu yd or less is frequently used. This standard requires an additional $\frac{1}{4}$ min of mixing for each cubic yard over 1 cu yd. Timing of mixing should not begin until all solid ingredients are in the drum; all of the water must be added before one-fourth of the mixing time has passed. If performance tests indicate that adequate mixing is obtained with shorter mixing times, some specifications permit the mixing time to be reduced. In practice, it has been found that the average time to produce a batch, including loading, mixing, and unloading, is about 2 to 3 min. Mixes having slumps of less than 4 in. may require a slightly longer mixing time at a slower mixing speed. Thus, single-drum mixer output may typically vary over the range of 15 to 35 batches per hour.

It is usually recommended that some water be added to the drum before the dry ingredients are added. This procedure helps to clean the drum and yields a more uniform mix. One suggested procedure involves adding 10% of the water to the drum before charging, 80% during charging, and the final 10% after all dry ingredients are in the drum. Admixtures should be added at the same point in each cycle if uniform results are to be obtained.

Estimating Mixer Production

The problem of mix design will not be considered here because this topic is covered in a number of references (see Reference 2). However, the method of determining the volume of concrete which a given mix will produce and a method for estimating mixer production are presented below.

When the quantity of ingredients (usually proportioned by weight) is specified, the volume of concrete produced by the mix can be found by summing the absolute volume of all components, including water and entrained air. This procedure is called the *absolute volume* method. The absolute volume of each component may be found by using the following equation:

$$\text{component volume} = \frac{\text{unit weight (SSD) (lb/cu ft)}}{62.4 \times \text{specific gravity}} \qquad (10\text{-}1)$$

Weights of aggregates specified in the mix design are usually based on saturated surface dry (SSD) conditions. That is, the aggregates will neither add nor subtract water from the mix. If free moisture is present in the aggregates, then a correction must be made to the quantity of water to be added at the mixer (see Example 10-1).

When small construction mixers are being used, the size of the batch will often be adjusted to utilize whole bags of cement (94 lb each) per batch. This procedure is illustrated in Example 10-1.

Once the volume of concrete to be produced in each batch has been determined, the hourly production of the mixer may be estimated by using Equation 10-2. In using this equation it is important that the batch cycle time be accurately estimated. As stated in Section 10-3, a minimum mixing time of 1 min is required for mixer volumes of 1 cu yd or less. However, experience has shown that batch cycle time usually averages between 2 and 3 min.

$$\text{production (cy/hr)} = \frac{2.22 \times V \times E}{T} \qquad (10\text{-}2)$$

where

$V =$ batch volume (cu ft);
$E =$ job efficiency;
$T =$ cycle time (minutes).

Example 10-1

PROBLEM—PART 1: Mix specifications call for the following ingredients per cubic yard of concrete:

Component	Quantity
Cement	526 lb
Sand (SSD)	1,438 lb
Gravel (SSD)	1,848 lb
Water	325 lb

Find the quantity of each ingredient to use in a 16S mixer; allow a 10% overload. Assume that aggregate is saturated surface dry.

Solution:

Batch volume $= 16 \times 1.10 = 17.6$ cu ft $= \dfrac{17.6}{27} = 0.652$ cu yd.

Cement $= 0.652 \times 526 = 343$ lb (3.65 bags).

Sand $= 0.652 \times 1{,}438 = 938$ lb.

Gravel $= 0.652 \times 1{,}848 = 1{,}205$ lb.

Water $= 0.652 \times 325 = 212$ lb.

PROBLEM—PART 2:   Since bagged cement will be used, you decide to use a three-bag mix. Find the quantity of each ingredient required per batch and the volume of concrete produced per batch.

Solution:

Using 3 instead of 3.65 bags of cement per batch will reduce all quantities by the factor $\dfrac{3}{3.65} = 0.822$.

Cement $= 3$ bags.

Sand $= 0.822 \times 938 = 771$ lb.

Gravel $= 0.822 \times 1{,}205 = 991$ lb.

Water $= 0.822 \times 212 = 174$ lb.

Batch volume $= 0.822 \times 17.6 = 14.5$ cu ft.

PROBLEM—PART 3:   You estimate a batch cycle time of 2 min and a job efficiency of 0.83. What is the estimated hourly production?

Solution:

$$P = \frac{2.22 \times 14.5 \times 0.83}{2.0} = 13.4 \text{ cu yd/hr (Equation 10-2)}.$$

PROBLEM—PART 4:   If the sand on the job contains 3% excess moisture and the gravel 1% excess moisture, what adjustments should be made to the three bag mix?

Solution:

Excess water in sand $= 771 \times 0.03 = 23$ lb.

Excess water in gravel $= 991 \times 0.01 = 10$ lb.

Total excess water $= 23 + 10 = 33$ lb.

New Mix Quantities:
   Water $= 174 - 33 = 141$ lb.
   Gravel $= 991 + 10 = 1001$ lb.
   Sand $= 771 + 23 = 794$ lb.

Ready-mixed Concrete

Ready-mixed concrete is concrete that is delivered to the job site ready for use. Such concrete may be mixed in a central mix plant before shipment or mixed in a transit-mix truck on the road or at the job site. Central-mix concrete may be delivered in nonagitating trucks, but it is commonly delivered in either a truck mixer operating at agitating speed or in an agitator truck.

A transit-mix truck used as an agitator truck can haul a volume of concrete larger than its rated mixing capacity. The manufacturer will provide the allowable capacity as an agitator. This capacity is usually about one-third greater than its rated mixing capacity. Specifications usually require transit-mix trucks to mix for 70 to 100 revolutions at mixing speed. All revolutions in excess of 100 should be at agitating speed. ASTM Standard C94-73a requires that truck-mixed concrete be discharged within $1\frac{1}{2}$ hr or before the drum has revolved 300 times, whichever occurs first, after the introduction of mixing water to the cement and aggregate or the introduction of the cement to the aggregate. When hauling long distances, transit-mix trucks would not add water and begin mixing until near the job site. A 12-cu yd transit mixer with retractable rear axle is shown in Figure 10-2.

Figure 10-2   Transit mixer with retractable rear axle. (Courtesy Challenge-Cook Bros., Inc.)

Paving Mixers

Although designed especially for paving operations, paving mixers are traveling mixers which operate basically in the same manner as the stationary mixers previously discussed. Dual drum mixers use two drums in series. Concrete is mixed for one-half the required time in the first drum and then it is transferred to the second drum to complete mixing. By this procedure production is almost double that of a single drum mixer.

A paving mixer is equipped with a boom and bucket for delivery of concrete to any desired point within the pavement forms. Sand, coarse aggregate (gravel or crushed stone), and cement are usually delivered in batch trucks to the loading skip of the mixer. Water is added from the mixer's storage tanks which are refilled by truck or pipeline.

Paving mixers are high-capacity mixing plants. The 34E dual drum mixer, for example, can produce about 75 cu yd/hr when operated as a traveling plant and 100 cu yd/hr when operated as a stationary plant.

Additional considerations in paving operations are discussed in Section 10-6.

10-4 TRANSPORTING EQUIPMENT

General

To maintain high quality concrete that has uniform properties, it is essential that all handling and transporting be carefully done from the mixer to the point of usage. Particular care must be taken so that the concrete will not segregate into its components. When concrete is being loaded from the mixer into trucks, buckets, etc., it is recommended that a downpipe having a vertical drop of at least 2 ft be used at the end of the chute or conveyor to prevent segregation. A similar arrangement should be used at the end of all chutes or conveyors used for transporting concrete.

Concrete may be transported by many different pieces of equipment. Some of these include wheelbarrows, buggies, chutes, conveyors, buckets, railroad cars, trucks, and pumps. Considerations involved in using each of these items are discussed in the following paragraphs.

Wheelbarrows and Buggies

Wheelbarrows are used primarily for transporting and placing small volumes of concrete. They can carry about $1\frac{1}{2}$ to $1\frac{3}{4}$ cu ft of concrete per trip. Wheeling planks or runways should be provided for both wheelbarrows and buggies. Pushing loaded wheelbarrows up wheeling planks should be avoided whenever possible.

Powered or pushed buggies carry larger volumes of concrete than wheelbarrows. Push buggies typically carry from 6 to 11 cu ft per load and can complete about 20 trips per hour on short, level runs. Powered buggies are available in capacities as large as $\frac{1}{2}$ cu yd and may operate efficiently on hauls up to 1,000 ft or so. The use of powered buggies requires careful attention to the design and construction of runways and attached forming. In addition to the static weight of the loaded buggies, impact and horizontal loads caused by braking of the buggies must be considered. At least one major collapse of a building under construction has been traced to buggy impact and braking loads on the supporting members of the formwork.

Chutes and Conveyors

Chutes are used primarily for moving concrete from the mixer to a transporter and for placing concrete into forms. It is recommended that metal or metal-lined chutes be used. The allowable slope of the chute will depend on the type of concrete involved. In general, a slope should be selected which will move the concrete at a sufficient speed to keep the chute clean but not fast enough to cause segregation. Downpipes should be used at the end of chutes. Conveyor belts may also be used for transporting concrete. Maximum slopes of $\frac{1}{2}$ in./ft upward or 1 in./ft downward are recommended for plastic concrete. Concrete conveyors should not be used to raise concrete more than about 30 ft. Concrete pumps or cranes with buckets should be used to raise concrete to higher elevations. Care must be used to avoid segregation at transfer and discharge points. Downpipes and hoppers may be used for this purpose. The belt should be wet or cleaned on its return to prevent a buildup of mortar on the belt, idlers, and supporting structure.

Buckets

Concrete buckets are manufactured in varying sizes and shapes; they typically hold from 1 to 8 cu yd. A gate and release mechanism are provided at the bottom of the bucket for unloading the concrete in the desired location. Although gates are often operated manually, remotely controlled power operated gates are safer when working above ground level. Buckets are commonly hoisted and moved by cranes but may be hauled on other equipment such as trucks and rail cars.

Trucks and Rail Cars

The use of transit-mix and agitator trucks for hauling concrete has already been mentioned. Such trucks are widely used because they cause minimum interference between concrete mixing and placing operations. They can also discharge at any desired rate over a relatively wide area using their own chutes. Dump trucks that have special concrete bodies are also sometimes used for hauling concrete. Their dump bodies are designed for ease of cleaning and dumping and for reducing segregation. Specifications may limit the time and speed at which plastic concrete may be hauled in nonagitator trucks depending on temperature and road smoothness. When a nonagitator truck is used for hauling concrete, segregation can be minimized by using a stiff mix with air entrainment.

Railroad cars especially designed for hauling concrete are sometimes used. Typical capacities range from 10 to 40 cu yd. Segregation problems and their solutions are similar to those of nonagitator trucks.

Pumps and Pipelines

Initially developed for moving concrete into tunnels, concrete pumping equipment is now widely used in building construction. Pumps are frequently mounted on a truck chassis and have elevating and turning mechanisms on their discharge lines to enable them to reach upper stories of low-rise buildings. Metal pipelines are commonly used with pumps, but flexible hoses may also be used. The maximum size of aggregate that can be pumped depends on the pump and pipeline size. Although it is possible to pump concrete of low slump, best results are usually obtained with air entrained concrete having a slump of 3 in. or more. When working with stiff mixes, it may be necessary to start the pumping operations with a thin concrete grout to lubricate the pipeline before concrete is introduced. A truck-mounted 70-cu yd/hr concrete pump with a placement boom capable of 110 ft vertical reach is shown in Figure 10-3.

Figure 10-3 Truck-mounted concrete pump and placement boom. (Courtesy Challenge-Cook Bros., Inc.)

Concrete pipeline and hose fittings are available that mount directly in concrete forms near their bottom. This permits filling the forms from the bottom rather than from the top. Such a procedure facilitates removal of air voids from the concrete in the forms and reduces the amount of vibration required during placing. However, use of this method increases the pressure on the form during placing.

10-5 PLACING, FINISHING, AND CURING

Placing

Placing concrete means to move plastic concrete into its final position (usually within forms) and includes spreading and consolidating. In placing, as in transportation, care must be taken to avoid segregation of the concrete. Before placing fresh concrete, check to see that the forms are clean, tight, and adequately supported. The interior surface of forms should be coated with form oil or some other parting agent to ease the removal of forms without damaging the hardened concrete's surface. Before the concrete is poured, the subgrade and any wood forms should be sealed or moistened to prevent their drawing water from the plastic concrete. Plastic sheets may be placed on the subgrade to provide a moisture barrier. Failure to adequately seal wood forms may result in wood sugar entering the concrete and retarding the hardening of the concrete.

If fresh concrete is to be placed on hardened concrete, preparation of the first (old) surface is required to achieve a satisfactory bond. This involves roughening the old surface either before or after it has hardened. Roughening before hardening may be accomplished by using water jets, brooms, or wire brushes. After hardening has taken place, roughening requires mechanical abrasions such as sandblasting, chipping, sawing, etc. A coating of grout or layer of mortar should also be placed on rock or hardened concrete before placing a layer of fresh concrete. This helps achieve a good bond.

During placing, concrete is deposited from the transportation equipment as close to its final position as possible. Concrete for slabs and massive structures can frequently be deposited directly from a bucket. Placing concrete into narrow vertical forms usually requires the use of chutes, hoppers, or pumps. The height of free fall of concrete may have to be limited to 5 ft or less if segregation occurs during placing. Downpipes or elephant trunks may also be required to prevent segregation.

Concrete should be placed in layers of from 6 to 24 in. in depth depending on the type of construction. The horizontal area of each layer must be limited to insure that the initial set does not take place before the next layer is poured.

Concrete may be placed under water by pumping or by using special underwater buckets or a *tremie*. The tremie consists of a hopper on top connected to a vertical pipe long enough to reach the desired location underwater. The hopper remains above the water. To prevent water from entering the tremie, the tremie is filled with concrete before it is lowered into position. When the bottom of the tremie reaches the desired elevation, concrete is allowed to flow out to an elevation above the end of the tremie. The tremie is slowly raised during pouring. Care must always be taken to keep the end immersed in plastic concrete. Concrete used with a tremie should have a water–cement ratio of less than $5\frac{1}{2}$ gal/sack, a cement content of at least 660 lb/cu yd, a slump of 6 to 7 in. and a maximum aggregate size of about 2 in.

Vibration and Consolidation

Concrete must be consolidated during placing to prevent the formation of void spaces (or honeycombing) in concrete. Consolidation may be accomplished by hand rodding or spading, but mechanical vibration is most effective. Electric or pneumatic immersion-type vibrators are commonly used. Form vibrators attached to the outside of forms are also sometimes used. The use of vibrators for consolidation permits the use of stiffer mixes having a higher coarse aggregate content than could be used with manual consolidation.

In vibrating concrete, vibrators should be inserted into the concrete vertically and allowed to penetrate to the bottom of the mix. Overvibration and the use of vibrators to move concrete horizontally must be avoided because these practices produce segregation. It is recommended that a vibrator be withdrawn and moved as soon as cement paste appears at the top of the vibrator.

Shotcrete

The term *shotcrete* has been adopted as standard by the American Concrete Institute to refer to concrete that is conveyed through a hose and pneumatically placed by projecting it onto a surface at high velocity. It includes both the dry mix process in which most of the water is added at the nozzle and the wet mix process in which all materials (including water) are mixed before entering the delivery hose. Other terms sometimes used for this type of process include *pneumatically applied concrete*, *gunite*, *sprayed concrete*, and *gunned concrete*. Because a relatively dry mix is used, the sprayed concrete can support its own weight and retain its shape even on vertical or overhead surfaces. This characteristic makes the process especially useful in the constuction of structures having compound curves such as swimming pools, tanks, tunnels, and canals. In addition to its structural use, it is also used for coating other materials including the fireproofing of structural steel.

Finishing

Finishing concrete involves bringing the surface to its final grade and imparting the desired surface texture. Operations include screeding, floating, troweling, and possibly brooming. Screeding involves striking off excess concrete to obtain the desired surface elevation. Floating serves to compact the surface as well as smooth the surface and embed aggregate particles. Troweling is used to produce a smooth, dense surface on the concrete. Brooming may be used to produce a textured skid-resistant surface.

Except for paving operations finishing principally involves hand tools. Power floats and trowels, however, are sometimes used. A three-unit riding-type power trowel is shown in Figure 10-4.

Figure 10-4   Triple unit power trowel. (Courtesy Master Division, Koehring Co.)

Curing

Proper curing of concrete requires that adequate moisture be available during hydration of the cement and that favorable temperatures be maintained. Common curing methods used to retain moisture include the use of wet coverings, ponding, sprinkling, paper or plastic sheets, and curing compounds. In recent years the use of sprayed-on curing compounds has become increasingly popular. Either hand or power operated spray equipment may be used.

10-6 CONCRETE PAVING

General

Because of the widespread use of concrete pavements for highways and airports, specialized paving equipment has been developed to permit their rapid and uniform construction. Until recent years, such equipment has primarily been designed to ride on the forms used to retain the concrete while it hardened and is thus referred to as *form-riding equipment*. The group of equipment used in paving is called a paving train. The increasing use of slipform paving in recent years has resulted in the development of specialized equipment for this process. Accessory equipment to perform such tasks as sawing may be used with either paving method.

Paving with Forms

Standard metal forms used for paving are typically 10 ft long and 8 to 12 in. in height. Forms are held in place by metal pins inserted through three pin sockets in each form and locked in place with locking wedges. A locking plate on one end and a receiving key on the other end are provided for the connection of adjacent forms. As mentioned earlier, the forms serve the dual purpose of retaining the concrete until it has hardened and of providing a track on which the other equipment rides.

To provide a uniform surface at the proper elevation for the subgrade, form-riding subgraders are available. These function in the same manner as the grade excavators and trimmers discussed in Chapter 8. However, they are designed to ride on the forms and use the forms for vertical control

The concrete spreader is a self-powered, form-riding unit that spreads, strikes off, and vibrates the concrete as it travels forward. Concrete is placed in the forms in front of the spreader by a paving mixer or by trucks from a central mix plant. When reinforcing steel is used, the paving mixer must operate outside the forms to avoid interfering with the placement of the reinforcing. Combination placer/spreader units are also available which can operate with either form or slipform paving methods. These units incorporate conveyor-hoppers that receive concrete from trucks and move it into position for spreading.

After the concrete has been spread, a transverse concrete finisher is used to shape the concrete to the desired cross section and to provide additional finishing. Next comes the longitudinal finisher which applies the final machine finish to the concrete.

The concrete finishing bridge is a form-riding platform that is used for hand finishing and serves as a bridge for crossing freshly poured concrete.

After all finishing operations are accomplished, a form-riding automatic curing machine is used to spray a curing compound uniformly over the fresh concrete. This unit is self-powered and equipped with a powered pump for spraying.

Slipform Paving

Slipform paving utilizes a slipform paver to spread, compact, and finish a concrete slab which is contained within the paver's moving forms. The concrete that has been placed is capable of supporting its own weight and retaining its shape without further support. The paver is designed to operate without interfering with the placement of reinforcing steel so that steel may be placed well ahead of the paver. A slipform paver capable of slipforming slabs with integral curb and gutter is shown in Figure 10-5.

Figure 10-5   Paver slipforming slab with integral curb and gutter. (Courtesy CMI Corp.)

Concrete used with a slipform paver must have a uniform consistency with a slump of 1 to 4 in. For maximum economy and strength, the concrete should be placed on a well-compacted and accurately graded base. Typical slipform paving machines can pour a slab up to 10 in. thick and 24 ft wide at speeds up to 20 fpm.

Other equipment that may be used with the slipform paver includes the grade trimming equipment discussed in Chapter 8, additional finishing equipment (such as a tube finisher or burlap drag), and curing machines.

Accessory Equipment

Additional equipment that may be used with either form or slipform paving includes concrete saws and portable curing machines. Concrete saws are used to cut longitudinal and transverse joints in finished slabs. Their crawling speed is adjustable to suit the hardness of the aggregate and the age of the concrete. Sawing is usually done from 6 to 30 hr after the concrete is placed. Trial cutting of a test section of concrete is recommended for determining the optimum time to saw joints. Sawing should be done as early as possible but not until a test cut produces clean unraveled edges. The usual depth of control joints is one-third to one-quarter of the slab thickness and not less than the maximum size of the aggregate used.

Portable curing machines are used to spray curing compound on concrete. Since they are manually controlled, they do not insure the uniform application rate of the automatic curing machine.

Curb, Gutter, Sidewalk, and Median Paving

In recent year there has been a rapid increase in the development of machines for rapid pouring of concrete curbs or curbs and gutters. Both slipform and extrusion-type machines are used, although slipform pavers predominate. Some machines operate as a combination grade trimmer and slipform paver to both prepare the grade and place the curb and gutter. Production claimed for these machines averages about 2,500 ft of finished curb and gutter per day with peak values of over a mile a day.

As an outgrowth of these curb and gutter paving machines, machines have been developed for slipforming sidewalk pavements and highway median barriers. Their production rate in feet per day is similar to that of curb and gutter machines. A small slipform paver capable of placing median barriers, curbs and gutters, sidewalks, and driveways is shown in Figure 10-6.

PROBLEMS

1. A one-sack trial mix that meets specification requirements has the proportions given below. Find the yield in cubic feet of this mix.

Component	Weight	Specific Gravity
Cement	94	3.15
Water	46	1.00
Sand (SSD)	183	2.65
Gravel (SSD)	362	2.66

Figure 10-6 Slipform paver for median barriers, curbs and gutters, etc. (Courtesy CMI Corp.)

2. How many integer one-sack batches of the mix in Problem 1 would be used to charge a 16S mixer if a 10% overload is allowed?

3. Find the quantity of each ingredient required for charging a 34E mixer with no overload using the mix proportions given below. Assume that surface moisture of 4% for fine aggregate and 1% for coarse aggregate will be present in the field.

Cement	94 lb
Fine aggregate (SSD)	200 lb
Coarse aggregate (SSD)	400 lb
Water	50 lb

4. Estimate the hourly production of a 34E dual drum mixer using a 10% overload, a cycle time of 1 min, and a job efficiency of 0.80.

5. What is the minimum mixing time required for a 4-cu yd mixer using the specification given in the text?

REFERENCES

1. *Concrete Construction*, Compilation No. 2. Detroit: American Concrete Institute, 1968.

2. *Design and Control of Concrete Mixtures* (11th ed.). Skokie, Illinois: Portland Cement Association, 1968.

11

BITUMINOUS EQUIPMENT

11-1 INTRODUCTION

General

Bituminous pavements and wearing surfaces provide a resilient, waterproof layer that protects the underlying materials from water and traffic. These pavements are often referred to as *flexible pavements* because of their ability to sustain small deflections caused by load or base course consolidation without damage. Properly designed bituminous pavements are less affected by temperature strains than are comparable rigid (concrete) pavements. Since bituminous pavements can be constructed from a wide range of materials, the use of available local materials often makes such pavements more economical than comparable rigid pavements. Bituminous pavements are particularly adaptable to stage construction where additional layers are added as traffic loads increase.

For the reasons cited above, bituminous surfaces and pavements have become widely used on highways, roads, and streets as well as airfields. This chapter will discuss the materials, equipment, and procedures used in constructing and maintaining such surfaces.

Materials

Bituminous surfaces consist of two principal parts: aggregates and a binder. The functions and desirable characteristics of each of these are described below.

Aggregates

Aggregates in a bituminous surface serve to transmit loads from the surface to the base course, resist the abrasive action of traffic, and provide a skid-resistant surface. Mineral aggregates such as broken stone and slag, crushed or uncrushed gravel, sand, and mineral filler are commonly used. Since aggregates usually make up 90 to 92% by weight of a bituminous mix, their properties greatly influence the characteristics of the bituminous surface. Some desirable characteristics of bituminous aggregates include the following:

1. Angular and rough, to increase their interlocking action

2. Hard and durable, to prevent cracking or crushing under load

3. Sound, to resist weathering

4. Properly graded, to permit obtaining the desired mix characteristics

5. Clean and dry, to permit proper bonding of the bitumimous binder

6. Hydrophobic (non water absorbing), to prevent stripping of the binder caused by water absorption in the aggregate

Aggregates are classified by size as coarse aggregate, fine aggregate, and fines (mineral filler or mineral dust). Coarse aggregate corresponds to gravel as described in Chapter 1, that is, does not pass the No. 4 sieve. Fine aggregate corresponds to sand, i.e., passes No. 4 sieve and is retained on No. 200 sieve. Fines (mineral filler) are any inert, nonplastic materials that pass the No. 200 sieve. Well-graded rock dust, portland cement, and hydrated lime are frequently used as fines in bituminous mixes.

Binder

The second component of a bituminous mix is the bituminous binder that binds the aggregate together and provides a waterproof cover for the underlying materials. Although all bituminous materials are black in color and are made up mainly of bitumen, they may be either asphalt or tar. Asphalt is derived from petroleum even though it may be found in natural deposits. Tars do not occur naturally but are manufactured from coal. Although most properties of asphalt and tar are similar, tars have the advantage of not being soluble in petroleum products. Thus, they are sometimes used on airfield areas where fuel spillage is likely to occur. Tars have the disadvantages of being subject to extreme consistency variation with moderate temperature changes.

Asphalt cement is a solid form of asphalt that must be heated to a liquid state for use in bituminous mixes. Asphalt cements are usually viscosity-graded based on their absolute viscosity measured at 140°F. Viscosity grades range from AC-2.5 (soft) to AC-40 (hard). A grading system based on a penetration test is also sometimes used. In this test the penetration (in hundredths of a centimeter) which occurs in 5 sec is measured for a standard needle under a 100-gram load with the asphalt at 77°F. Penetration grades range from soft (penetration numbers 200 to 300) to hard (penetration numbers 40 to 50).

Asphalt cement diluted with petroleum distillates yields a product that is liquid at room temperature and is referred to as an *asphalt cutback*. Cutbacks are classified as rapid curing (RC), medium curing (MC), or slow curing (SC) and are assigned a number that represents their viscosity grade (usually measured as kinematic viscosity). Viscosity grades range from 30 (viscosity similar to water) to 3,000 (barely deforms under its own weight). Another (older) grading system assigns viscosity grade numbers 0 (low viscosity) through 7 (high viscosity). Asphalt emulsions are available that can be diluted with water and are nonflammable. Tars are also available in a range of twelve viscosities that varies from very fluid (RT-1) to solid (RT-12) at room temperature and as cutbacks.

Precautions in Handling Bituminous Materials

The flash point of a bituminous material is the temperature at which an instantaneous flash will occur in the presence of an open flame. The flash point is usually well below the temperature at which the material will burn.

When bituminous materials are heated for spraying or mixing, it must be recognized that the temperatures employed are usually above the flash point for bituminous cutbacks. Thus, adequate fire safety precautions must be taken to prevent injury and/or equipment damage. No spark producing equipment or open flame should be permitted near the bituminous materials. Equipment used for storing, heating, mixing, and spreading of bituminous materials must have been designed and approved for such use. Personnel handling cutbacks must be trained in the proper method of handling these materials and must be adequately supervised. Fire extinguishing equipment should be readily available for use if needed.

Types of Bituminous Surfaces

The types of bituminous surfaces that may be applied range from a simple spray coating (dust palliatives and preliminary treatments) to the plant-mixed, high-type bituminous pavement. Single and multiple surface treatments, seal coats, and penetration macadam are discussed in Section 11-2 and road and plant mix pavements are covered in Section 11-3.

Bituminous dust palliatives use a spray of bituminous material (usually applied by the bituminous distributor described below) to penetrate and coat soil particles in the surface of unpaved roads in order to reduce dust and provide some waterproofing of the surface. Bituminous materials frequently used are liquid asphalts (such as SC-70, MC-30, or MC-70) or diluted slow-setting emulsions. Road oiling is similar to applying dust palliatives except that the term usually denotes a part of a continuing buildup of the road surface over a period of years.

Preliminary treatments are used to provide a bond between an existing surface and a new bituminous surface. Preliminary treatments include prime coats and tack coats. Prime coats are applied to a porous, nonpaved surface. Priming material is usually a low-viscosity asphalt (such as MC-30, MC-70, or MC-250), a low-viscosity tar, or diluted asphalt emulsion. The amount applied should seal the surface, plug capillary voids, and coat and bond loose mineral particles. It should cure in about 48 hr. Usual quantities applied range from 0.15 to 0.40 gal/sq yd.

A tack coat is applied to an existing paved surface. It should be applied uniformly in a very thin coat. Usual quantities required range from 0.05 to 0.25 gal/sq yd. Materials used include asphalt cutbacks (RC-250 or RC-800), road tars (RT6-11), a diluted emulsion, or an asphalt cement (AP-1 or AP-3).

The Bituminous Distributor

The piece of equipment used in virtually all bituminous surface construction is the bituminous (or asphalt) distributor shown in Figure 11-1. It is used for applying dust palliatives and preliminary treatments as well as in constructing the surface treatments, seal coats, and macadam pavements described in Section 11-2.

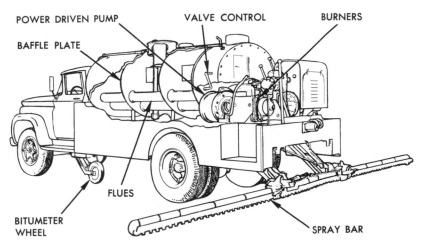

Figure 11-1 Typical bituminous distributor. (Courtesy The Asphalt Institute.)

The rate of application (normally expressed in gallons per square yard) of bituminous material by a bituminous distributor is controlled by three variables: pump speed, distributor road speed, and spray bar length. Typical spray bars are adjustable from 4 to 24 ft in length. Pump output is indicated by a tachometer, usually calibrated in gallons per minute. Distributor road speed is measured by a bitumeter, usually calibrated in feet per minute. Since the volume of asphalt varies with temperature, most specifications are based on the volume at 60°F. Thus, a temperature correction factor must be applied to volume measured at other temperatures in order to convert to the standard volume.

For a particular spray bar length, the road speed (bitumeter reading), and pump output (tachometer reading) needed to obtain a specified application rate can be found in the tachometer chart supplied by the distributor manufacturer. If a tachometer chart is not available, the necessary road speed (bitumeter reading) can be found by using Equation 11-1.

$$S = \frac{9 \times P}{W \times R} \qquad (11\text{-}1)$$

where
S = road speed (fpm);
P = pump output (gpm);
W = bar width (ft);
R = application rate (gal/sq yd).

Example 11-1

PROBLEM: Find the distributor speed necessary to obtain an application rate of 0.20 gal/sq yd if spray bar length is 20 ft and pump output is 200 gpm.

Solution:

$$S = \frac{9 \times 200}{20 \times 0.20} = 450 \text{ fpm (Equation 11-1).}$$

The length of spread that one distributor load of asphalt will provide can be calculated by the use of Equation 11-2. If the distributor is not equipped with a sump, approximately 50 gal of asphalt should be left in the tank to prevent an uneven application rate at the end of the spread caused by air entering the pump suction line. The spread rate should be checked after each application of asphalt. This can be done by measuring the volume of asphalt applied by the distributor (volume at the start minus volume after spreading), converting this to the volume at standard temperature, and then solving Equation 11-2 for R (application rate).

$$L = \frac{9 \times Q}{W \times R} \qquad (11\text{-}2)$$

where
L = length of spread (ft);
Q = quantity of asphalt (gal);
W = width of spray (ft);
R = application rate (gal/sq yd).

Example 11-2

PROBLEM: Find the length of spread for the distributor in Example 11-1 if the tank holds 2,000 gal. The tank has a sump.

Solution:

$$L = \frac{9 \times 2,000}{20 \times 0.20} = 4,500 \text{ ft (Equation 11-2).}$$

It is important that the distributor apply the bituminous material uniformly over the surface to be treated. The following procedures are suggested to insure uniform application:

1. Use the nozzle size recommended by the distributor manufacturer for the particular application. Be sure that all nozzles are the same size.

2. Set all nozzles at the angle with the spray bar recommended by the distributor manufacturer (usually 15° to 30° from the spray bar axis). Manufacturers usually provide special wrenches for this purpose to insure uniform adjustment. Do not permit operators to set the end nozzles at a different angle in an attempt to get a clean edge. This practice will result in an excessive rate of application along the edge. Use an end curtain or special end nozzle to get a clean edge if desired.

3. Set the spray bar at the height above the road surface recommended by the distributer manufacturer. This should produce a uniform spray with two or three spray patterns overlapping (double or triple overlap). Check the spray bar height at the beginning and end of each spread.

4. Perform a spray test before beginning each day's application to see that all sprays are working properly (unclogged) and providing uniform coverage. Check that the cutoff valve provides rapid opening and cutoff operation.

5. While spraying insure that road speed and pump output are maintained at the specified rate. Spraying pressure must be high enough to give clean spray fans at each nozzle.

11-2 SURFACE TREATMENTS, MACADAM, AND ROAD MIXES

General

The term bituminous surface treatment covers a wide range of bituminous applications used to improve the surface of a road. Such treatments may involve the use of bituminous material alone (such as dust palliatives and preliminary treatments) or bituminous-aggregate combinations (single and multiple surface treatments, etc.). However, the term surface treatment is usually limited to a product thickness of less than 1 in. Thicknesses of 1 in or more are referred to as *pavements*. Road mixes, for example, may be used to produce either surface treatments or pavements. Although penetration macadams are not surface treatments, their construction is similar to that of surface treatments and thus they are included in this section. The construction of road and plant mixed pavements is covered in Section 11-3.

Seal Coats

A seal coat is a thin surface treatment applied to improve the surface texture and to waterproof an existing surface treatment or pavement. Only the fog seal is applied without any aggregate. A fog seal consists of a light application (0.1 to 0.2 gal/sq yd) by a bituminous distribution of a slow-setting asphalt emulsion diluted with an equal quantity of water. It seals small cracks and voids and renews old asphalt surfaces.

An emulsion slurry seal utilizes a combination of slow-setting emulsified asphalt, fine aggregate, mineral filler, and water, mixed to a slurry consistency. It is used to rejuvenate old pavements and to fill large cracks and spalled areas. Usual combinations (by weight of dry aggregates) provide for 20 to 25% asphalt emulsion and 10 to 15% water. Mineral filler usually makes up 5 to 15% of the total weight of aggregate and mineral filler. Usual applications range from $\frac{1}{16}$ to $\frac{1}{8}$ in. in thickness but should not exceed $\frac{1}{4}$ in. for a single course. Emulsion slurry seals may be dumped onto a surface and spread by hand-operated squeeges. More often, they are mixed in transit-mix trucks and spread by a spreader box towed by the transit-mix truck. Slurry seal machines are also available which both mix and spread the material in a single operation. The slurry seal machine consists of aggregate, asphalt emulsion and water storage containers, a mixer section, and a spreader section mounted on a truck chassis. A calibrated feeder gate controls the flow of aggregate onto a conveyor belt which feeds the mixer. Water and asphalt emulsion is metered and fed directly into the mixer. After being mixed, the slurry is fed into an adjustable spreader box. Control procedures for the unit are similar to those for other mixing equipment. Items to be checked include: proper calibration of aggregate feeder gate and metering pumps, proper operation of agitator, and proper adjustment of screen height and width.

A sand seal consists of an application of bituminous material covered with fine aggregate. It is used to provide skid resistance and to waterproof the surface.

Single and Multiple Surface Treatments

Single and multiple surface treatments, sometimes called *aggregate surface treatments*, are made up of alternate applications of asphalt and aggregate. A *single surface treatment* consists of a sprayed bituminous material covered by an aggregate layer approximately one stone in depth. Thus, the thickness of the surface treatment is about the same as the maximum aggregate size used. A multiple surface treatment consists of two or more surface treatments placed on top of each other. These are referred to as *double surface treatments* (two layers) or *triple surface treatments* (three layers). The maximum size aggregate used in each layer of a multiple surface treatment should be about one-half the size used in the previous layer. A multiple surface treatment produces a dense, waterproof, wearing surface that also adds some strength to the underlying material.

Since the construction of aggregate surface treatments requires a minimum of time, equipment, and material, these surfaces are widely used. Such surface treatments also lend themselves to stage construction as traffic loads increase. It is important that surface treatments be used with a base of adequate strength so that excessive repair and maintenance may be avoided. When a surface treatment is to be applied over an existing sand-clay or clay-gravel surface, the plasticity index of the existing surface material should be carefully checked to prevent future problems caused by softening of the base by moisture.

The basic sequence of constructing a single surface treatment consists of the following steps:

1. Sweeping

2. Prime application and curing (when required)

3. Binder application

4. Aggregate application

5. Rolling

6. Sweeping

When multiple surface treatments are being constructed, steps 3, 4, and 5 are repeated for each course.

Before construction is begun, the existing surface should be carefully checked and repaired as necessary. Granular bases and unpaved roads should be shaped to the required grade and cross section, and any soft spots should be removed and replaced with material of adequate strength. Old pavements should be repaired and cracks should be filled. The surface should then be thoroughly cleaned.

The equipment used for cleaning the surface usually is the rotary power broom, but blowers, flushing trucks, and hand tools may also be used. Rotary brooms may be self-propelled or towed. Cleaning equipment should be checked for proper operation and bristle wear before being used.

After the surface is swept, a prime coat is applied to untreated surfaces and is allowed to cure for at least 24 hr. If the bituminous material in a prime coat is not completely absorbed by the base at the end of the curing period, just enough sand to absorb the excess bituminous material should be applied. The piece of equipment used for bituminous application is, of course, the bituminous distributor. In addition to the general precautions and adjustments previously mentioned, care must be taken to insure that the bituminous material is applied at the required temperature and rate. For transverse joints, heavy paper should be used to provide a clean cutoff at the beginning and end of each spread. This will prevent a buildup of bituminous and aggregate at the joints.

The spreading of aggregate must follow immediately after binder application. Tests have shown that binder temperature usually drops to the road surface temperature within 1 min after binder application. Hence, aggregate should be applied within 1 min after binder application to insure proper bonding of binder and aggregate. The piece of equipment used for spreading aggregate is the aggregate spreader. There are a number of types of aggregate spreaders, some of which are illustrated in Figure 11-2. The whirl-type spreader [Figure 11-2(a)] may be built on or attached to an aggregate truck. It is used for applying fine aggregates. The vane spreader [Figure 11-2(b)] attaches to the tailgate of a truck and is thus classified as a tailgate spreader. Another form of tailgate spreader uses a hopper attached to the tailgate. Mechanical spreaders mounted on their own wheels and pushed by a dump truck include the hopper spreader shown in Figure 11-2(c). Self-propelled spreaders which provide a continuous and very uniform application of aggregate are also available. The flow of aggregate through such a spreader is illustrated in Figure 11-2(d).

It is essential to insure that aggregate spreaders are adjusted to provide a uniform spread of aggregate at the desired rate. Calibration may be checked by measuring the area covered by one truck load of aggregate and then dividing the weight of aggregate used by the area covered. (The rate of aggregate application is normally specified in pounds per square yard). Calibration may also be checked by weighing the amount of aggregate deposited on a square yard of cloth, paper, or box which is placed in the path of the spreader. Drag brooms are sometimes used to redistribute aggregate that has not been uniformly applied. However, drag brooming tends to displace aggregate from the binder and should be avoided whenever possible. Properly applied aggregate should not require drag brooming.

In an attempt to overcome the problems involved in applying aggregate within 1 min of binder application, single pass surface treatment machines have been developed. These machines apply both binder and aggregate in a continuous operation. Such machines may be either self-propelled, mounted on a motor grader, or towed by dump trucks. Calibration, adjustment, and operation procedures are basically the same as those for bituminous distributors and aggregate spreaders.

After the aggregate is applied, the surface must be rolled to embed the aggregate in binder and interlock the aggregate particles. Either pneumatic or steel-wheeled rollers may be used for this purpose. However, pneumatic rollers are recommended because they reduce bridging action during rolling and their contact pressure can be easily varied to prevent crushing of the aggregate particles. All rollers should be checked to insure that they are in good mechanical condition before use. Items to be checked include smooth operation in starting, stopping or reversing, condition of tires and treads, wear of rims on steel-wheeled rollers, condition of wheel and steering bearings, etc. Roller weight and tire pressure must be adjusted to yield the desired contact pressure. After rolling is completed and the binder has cured, any excess (unbonded) aggregate should be removed by lightly sweeping with a power broom. This will prevent damage caused by fast vehicles throwing stones.

Other pieces of equipment that may be used in constructing asphalt surface treatments include heater planers and repavers. These machines are used to level and reshape old asphalt surfaces prior to placing a new surface treatment. The

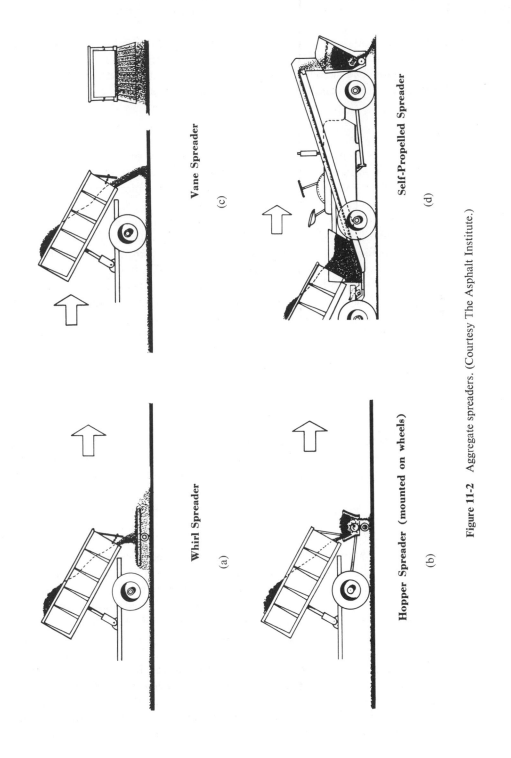

Whirl Spreader

(a)

Hopper Spreader (mounted on wheels)

(b)

Vane Spreader

(c)

Self-Propelled Spreader

(d)

Figure 11-2 Aggregate spreaders. (Courtesy The Asphalt Institute.)

195

heater planer heats and planes the old surface to the desired contour; the repaver adds and compacts any new material required for leveling. Items to be checked on such equipment include: condition and adjustment of cutter blades, operation of burners, and the condition of the compaction rollers on the repaver.

Penetration Macadam

Bituminous penetration macadam is not a form of surface treatment, but the equipment and construction procedures used are essentially the same as those used for aggregate surface treatments. Penetration macadam may be used as base courses or as pavements.

Construction of penetration macadam involves placing a single layer of coarse aggregate and rolling (keying) it with a steel-wheeled or vibratory roller. Bituminous material is applied and is then immediately covered with an application of intermediate size aggregate and compacted. If the macadam is to be used as a base, a tack coat is applied to the macadam surface prior to placing the pavement. If the macadam is to be used as a pavement, a single or multiple surface treatment is placed on top of the macadam.

11-3 BITUMINOUS PAVEMENTS

General

As mentioned previously, bituminous pavements are generally considered to be bituminous surfaces 1 in. or more in thickness. Pavements are constructed from road mixes (mixed-in-place construction) or plant mixes. Bituminous plant mixes may be either hot or cold mixes which are produced in stationary or portable plants.

Road Mixes

Road-mix or mixed-in-place construction utilizes aggregate already available on the road or taken from nearby locations and mixes it with bituminous material. This bituminous mix is then spread and compacted to form a pavement. The principal advantages of this method of construction include speed, economy, and a minimum investment in equipment. Disadvantages include limited control of mix quality as well as dependence on weather for control of moisture in the aggregate. The traveling plants described below provide improved quality control (possibly including a dryer) at a considerably larger investment in equipment. The traveling plant therefore falls in a category between the usual road-mix equipment and a stationary plant in both quality of product and equipment investment.

Road mixes may be produced by graders and bituminous distributors, rotary mixers, or travel plants. Rotary mixers combine aggregate with bituminous material by using a pulverizer-mixing rotor that is frequently equipped with a spray system for adding bituminous and/or water. Travel plants receive aggregate from trucks or pick it up from a windrow, mix it with bituminous, and either deposit the mix in a windrow behind the machine or place the mix in a manner similar to that of the asphalt paver.

Equipment that may be used with a road mix includes motor graders, bituminous distributors, water distributors, dump trucks, and rotary brooms. Travel plant operations may also require the use of asphalt tank trucks and pavers. Major problems involved in road mix construction usually include the control of aggregate moisture content, uniform mixing of aggregate and binder, and uniform spreading of the mix.

Plant Mixes

Hot-mix bituminous concrete pavement is considered the highest type of flexible pavement. It is suitable for use on airports as well as highways and streets. Since they require no curing, hot-mix pavements may be used as soon as the pavement has cooled to the ambient temperature. After compaction and cooling, such pavements are very stable and resist damage caused by moisture or frost.

Cold-mix pavements are constructed in generally the same manner as are hot-mix pavements. They have certain advantages in that they can be transported long distances, stockpiled if necessary, and used in only the quantity needed. However, they have the disadvantages of requiring curing, having low initial stability, and being difficult to compact adequately in cold weather. Since hot mixes predominate in flexible pavement construction, only their construction will be discussed here.

Paving operations frequently require large quantities per hour of asphalt mix. For example, an asphalt paver laying a 3-in. thick pavement 12 ft wide at a speed of 50 fpm requires 600 ton/hr of hot mix. Thus, storage, handling, and hauling of the plant mix must be carefully planned and controlled. The use of insulated tanks such as shown in Figure 11-3 for storing the hot mix at the job site may be necessary when hauling capacity is limited or is uncertain because of traffic conditions, etc. A special asphalt hauling unit is shown in Figure 11-4.

Spreading the Mix

Constructing hot-mix bituminous pavements primarily involves spreading and compacting the mix. The machine commonly used for spreading the mix is the asphalt paver or finishing machine. Figure 11-5 illustrates a typical paver in operation. The two principal parts of the paver are the tractor unit that propels the paver, pushes the dump truck delivering the mix, and pulls the screed unit and the screed unit that strikes off the mix at the proper level and provides initial compaction to the pavement. Most pavers now provide an automatic control system that uses a fixed string-line, ski, shoe, or traveling string-line as an elevation

Figure 11-3   Hot mix asphalt storage bins. (Courtesy CMI Corp.)

reference to automatically control the screed elevation. Towed-type pavers are also available for use on small paving jobs. The flow of material through this type of paver is illustrated in Figure 11-6. Another type of paver is the shoulder paver. This is a small paver (maximum paving width about 10 ft) used for paving highway shoulders or for widening existing pavements. These machines are available as attachments for a motor grader or as self-propelled machines.

Bituminous paving mixes are also spread by motor graders or spreader boxes. When graders are used for spreading, the mix is usually deposited from dump trucks into windrows using windrow sizers or eveners to provide a uniform windrow. The mix is then spread by the grader blade in the same manner as a road mix. The use of graders with automatic blade control devices will greatly improve the uniformity of the surface elevation. Spreader boxes towed by dump trucks operate in a manner similar to aggregate spreaders. The use of spreader boxes for paving is normally confined to small paving jobs.

Items to be checked during paver operation include grade and tolerance of the finished surface, appearance and temperature of the mix, weight of mix applied per square yard, and average thickness of mix actually obtained.

Steel-wheeled rollers should roll with the drive wheel forward, particularly during initial rolling, to prevent the displacement of the mix as illustrated in Figure 11-7. The sequence of rolling following spreading of the mix should be as follows:

Figure 11-4   Truck for hauling hot mix asphalt. (Courtesy J. H. Holland Co.)

1. Roll any transverse joints

2. Roll longitudinal joints

3. Roll outside edge

4. Breakdown rolling

5. Intermediate rolling

6. Finish rolling

Transverse joints, longitudinal joints, and outside edges are usually rolled with two-axle, tandem rollers. Recommended procedures for rolling transverse and longitudinal joints are illustrated in Figure 11-8.

The use of 10- to 12-ton three-wheeled steel rollers is suggested for breakdown rolling, but tandem steel-wheeled rollers may also be used. Pneumatic-tired rollers are recommended for intermediate rolling because they provide a more uniform contact pressure than do steel-wheeled rollers and they improve the amount of surface sealing obtained during rolling. Three-axle tandem steel-wheeled rollers are recommended for finish rolling because of their leveling action (illustrated in Figure 11-9). Two-axle tandem rollers may also be used for finish

Figure 11-5 Asphalt paver in operation. (Courtesy CMI Corp.)

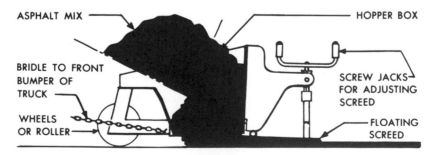

Figure 11-6 Towed-type paver. (Courtesy The Asphalt Institute.)

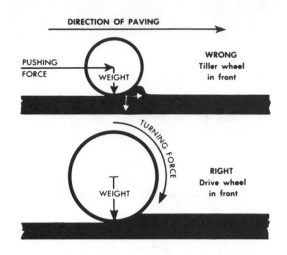

DIRECTION OF PAVING

PUSHING FORCE

WEIGHT

WRONG
Tiller wheel in front

TURNING FORCE

WEIGHT

RIGHT
Drive wheel in front

Figure 11-7 Direction of rolling pavement. (Courtesy The Asphalt Institute.)

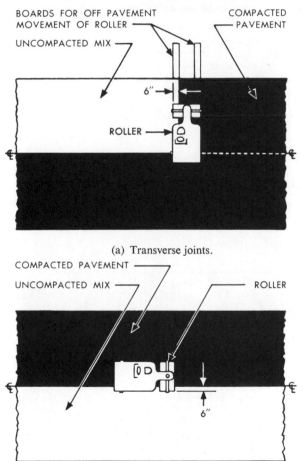

BOARDS FOR OFF PAVEMENT MOVEMENT OF ROLLER

COMPACTED PAVEMENT

UNCOMPACTED MIX

6"

ROLLER

(a) Transverse joints.

COMPACTED PAVEMENT

UNCOMPACTED MIX

ROLLER

6"

(b) Longitudinal joints.

Figure 11-8 Proper method of rolling joints. (Courtesy The Asphalt Institute.)

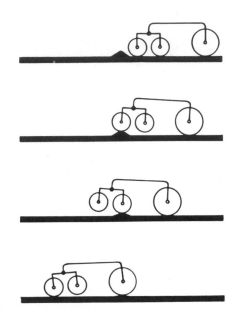

Figure 11-9   Leveling action of three-axle tandem roller. (Courtesy The Asphalt Institute.)

TABLE 11-1   *Possible causes of imperfections in finished pavement.*
(Courtesy The Asphalt Institute.)

Insufficient or Non-Uniform Tack Coat	Improperly Cured Prime or Tack Coat	Mixture Too Coarse	Excess Fines in Mixture	Insufficient Asphalt	Excess Asphalt	Improperly Proportioned Mixture	Unsatisfactory Batches in Load	Excess Moisture in Mixture	Mixture Too Hot or Burned	Mixture Too Cold	Poor Spreader Operation	Spreader in Poor Condition	Inadequate Rolling	Rolling at Wrong Time	Over-Rolling	Rolling Mixture When Too Hot	Rolling Mixture When Too Cold	Roller Standing on Hot Pavement	Overweight Rollers	Roller Vibration	Unstable Base Course	Excessive Moisture in Subsoil	Excessive Prime Coat or Tack Coat	Poor Handwork Behind Spreader	Excessive Hand Raking	Labor Careless or Unskilled	Excessive Segregation in Laying	Faulty Allowance for Compaction	Operating Finishing Machine Too Fast	Mix Laid in Too Thick Course	Traffic Put On Mix While Too Hot	Types of Pavement Imperfections That May Be Encountered In Laying Plant Mix Paving Mixtures.
				x	x	x															x											Bleeding
					x					x	x																					Brown, Dead Appearance
				x	x	x															x											Rich or Fat Spots
		x	x		x	x			x	x	x	x	x	x	x	x	x							x	x	x	x		x			Poor Surface Texture
x	x	x			x	x			x	x	x	x				x	x	x	x	x	x			x	x	x	x		x			Rough Uneven Surface
		x		x		x	x		x	x	x	x				x				x				x	x	x	x					Honeycomb or Raveling
				x					x	x	x	x	x		x	x		x	x	x	x			x	x	x	x	x				Uneven Joints
				x		x	x					x		x	x				x	x	x	x				x						Roller Marks
x	x			x		x	x	x	x				x	x		x					x	x	x				x			x	x	Pushing or Waves
				x	x	x											x	x	x		x		x	x								Cracking (Many Fine Cracks)
																	x				x		x	x								Cracking (Large Long Cracks)
		x		x		x										x	x	x			x											Rocks Broken by Roller
		x		x		x		x	x	x	x		x	x	x	x					x					x			x			Tearing of Surface During Laying
x	x		x		x	x		x		x			x			x	x	x		x	x	x	x									Surface Slipping on Base

rolling. The use of vibratory and combination conventional/vibratory rollers in bituminous pavement is increasing.

Before rolling operations are begun, roller condition must be checked as described previously under surface treatment construction. Items to be checked during rolling include the adequacy of compaction, surface smoothness, the use of proper rolling patterns and procedures, and the condition of joints and edges. A number of possible causes for imperfections in finished pavements are indicated in Table 11-1.

11-4 ASPHALT PLANTS

General

The plant that produces the bituminous mix for hot-mix paving is commonly referred to as an *asphalt plant*. There are three major types of asphalt plant: the batch plant, the continuous mix plant, and the drum mix plant. In a batch plant, accurately weighed portions of hot aggregate are placed into a mixing chamber (pugmill), the required quantity of asphalt is added, and materials are mixed and discharged into trucks or hot-mix storage bins. In a continuous mix plant, control devices are calibrated so that properly proportioned quantities of aggregate and asphalt flow into the pugmill in a continuous stream. Mixing occurs as the material flows through the pugmill so that mixing is completed by the time the material is discharged at the end of the pugmill. The major components of a continuous flow plant and the flow of materials through the plant are illustrated in Figure 11-10.

A drum mix plant is similar to the continuous flow plant except that both aggregate drying and mixing occur in the dryer drum. Thus, neither hot elevator, nor gradation control unit, nor pugmill is required. Stack emissions from the dryer are also greatly reduced since the asphalt tends to trap the fines in the drum, greatly reducing the need for pollution control equipment. Because of these differences, drum mix plants have lower initial costs as well as lower operating and maintenance costs than conventional plants. However, they may not provide as high a quality mix as a conventional plant. The components of a drum mix plant are shown in Figure 11-11.

Operation of a continuous flow plant takes place as follows. Cold aggregate is placed into the bins of the cold feed hopper from stockpiles. Since several different sizes of aggregates are usually combined to produce a mix that meets gradation specifications, several bins are used to hold the different size aggregates. Aggregates are released from the cold feed unit at a controlled rate and are fed by the cold elevator to the dryer where moisture is removed and the aggregates are heated. The hot elevator then takes the heated aggregate to the gradation control unit (GCU) where the hot aggregate is separated (by means of screens) into bins by size. Calibrated feeder gates in the gradation control unit and in the fines feeder unit control the flow of the hot aggregates and mineral filter to a second hot

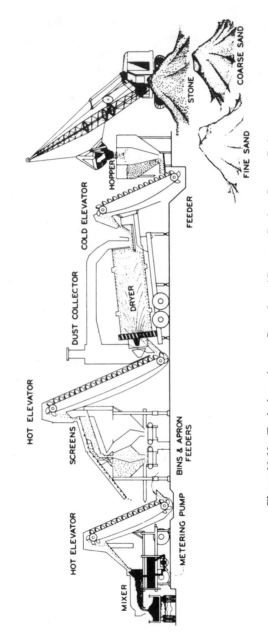

Figure 11-10 Typical continuous flow plant. (Courtesy Barber-Greene Co.)

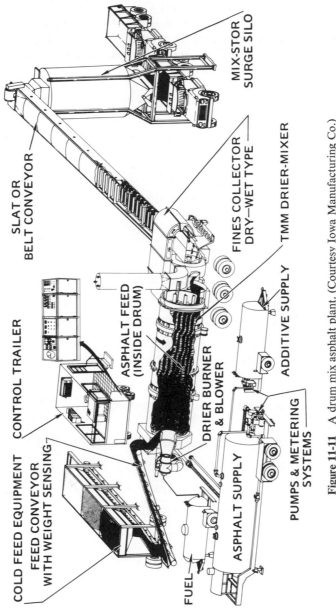

COLD FEED EQUIPMENT

CONTROL TRAILER

FEED CONVEYOR
WITH WEIGHT SENSING

SLAT OR
BELT CONVEYOR

MIX-STOR
SURGE SILO

ASPHALT FEED
(INSIDE DRUM)

FINES COLLECTOR
DRY—WET TYPE

TMM DRIER-MIXER

DRIER BURNER
& BLOWER

ADDITIVE SUPPLY

ASPHALT SUPPLY

PUMPS & METERING
SYSTEMS

FUEL

Figure 11-11 A drum mix asphalt plant. (Courtesy Iowa Manufacturing Co.)

elevator which carries them into the pugmill. Metering pumps release a controlled flow of asphalt into the mix as it enters the mixer. As the mix moves through the mixer, mixing is completed and the finished mix is discharged into haul units.

To produce a continuous flow of hot-mix asphalt of uniform quality requires that all plant components be carefully calibrated as described in the following subsection. Automatic control and recording devices are now available on many plants to facilitate quality control of the plant product.

Continuous Flow Plant Calibration

In order to calibrate a continuous flow plant, the following information must be known:

1. Mix specifications
   (a) Percent and type of bitumen
   (b) Blended aggregate gradation

2. Number and capacity of GCU bins to be used

3. Screen sizes available

4. Aggregate moisture content

A recommended sequence of operations for plant calibration is given below. These procedures are illustrated in Example 11-3.

1. Select screen sizes to be used in the gradation control unit based on bin capacities, gradation of the blended aggregate, and screen sizes available.

2. Determine dryer capacity.

3. Determine plant output based on dryer capacity and selected aggregate gradation.

4. Determine the feeding rate for aggregates and asphalt.

5. Determine GCU feeder speeds.

6. Calibrate GCU feeder gates.

7. Calibrate the fines (mineral filler) feeder.

8. Calibrate the asphalt metering pump.

9. Determine initial cold feed gate settings.

10. Check all settings.

11. Produce the mix on a trial basis.

Example 11-3

PROBLEM: Perform the initial calibration and calculate the production of an asphalt plant based on the data below.

(a) Mix specification:
  Bitumen content: 6% of AC 85-100.
  Aggregate gradation: Figure 11-12.

(b) Number of GCU bins = 4

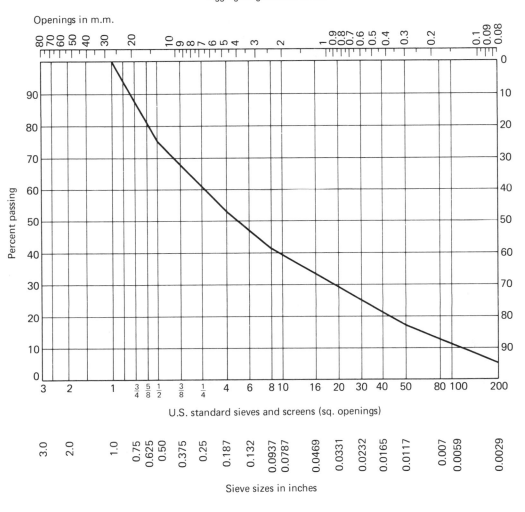

Figure 11-12 Aggregate blend—Example 11-3.

(c) Screen sizes available: Figure 11-13.

(d) Aggregate moisture content $= 7\%$. (See Figure 11-13 for dryer capacity).

(e) Gradation unit bin capacities:

$$
\begin{aligned}
\text{Bin } \#1 &= 13.5 \text{ tons} \\
\text{Bin } \#2 &= 0.6 \text{ tons} \\
\text{Bin } \#3 &= 4.5 \text{ tons} \\
\text{Bin } \#4 &= \underline{0.6} \text{ tons} \\
\text{Total} &= \underline{30} \text{tons}
\end{aligned}
$$

Dryer capacity chart

Capacity (tons per hour)

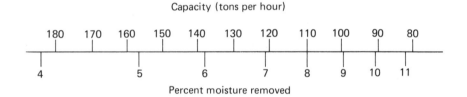

Screen data

Material size (sieve size)	Screen size required to pass material (in.)
#8	$\frac{1}{10}$
#6	$\frac{5}{32}$
#4	$\frac{1}{4}$
$\frac{5}{16}''$	$\frac{3}{8}$
$\frac{1}{2}''$	$\frac{9}{16}$
$\frac{3}{4}''$	$\frac{7}{8}$
$1''$	$1\,\frac{1}{8}$
$1\,\frac{1}{4}''$	$1\,\frac{3}{8}$
$1\,\frac{1}{2}''$	$1\,\frac{5}{8}$

Figure 11-13 Dryer and screen data—Example 11-3.

Solution:

(a) Select screen sizes for gradation control unit: Find percent of total capacity available in each bin:

$$\text{Bin } \#1 = \frac{13.5}{30} = 45\% \qquad \text{Bin } \#3 = \frac{4.5}{30} = 15\%$$

$$\text{Bin } \#2 = \frac{6}{30} = 20\% \qquad \text{Bin } \#4 = \frac{6}{30} = 20\%$$

Using Figure 11-12, select material size ranges that will divide aggregate among the bins in proportion to bin capacity. Note in Figure 11-13 that the finest screen available will pass #8 material and smaller. Bin size separations are thus chosen as follows:

Bin #	Material Sizes	% of Total
1	#6–#200	42*
2	$\frac{5}{16}$ in.–#6	18
3	$\frac{1}{2}$ in.–$\frac{5}{16}$ in.	10
4	1 in.–$\frac{1}{2}$ in.	25

*47% − 5% mineral filler (#200 and smaller) = 42%.

Corresponding screen sizes selected are thus $1\frac{1}{8}$ in., 9/16 in., 3/8 in., 5/32 in. (Figure 11-13). Aggregates are then divided by screens and stored as follows:

Passing Screen Size (in.)	Retained on Screen Size (in.)	Storage (Bin #)	Percent of Total
$1\frac{1}{8}$	$\frac{9}{16}$	4	25
$\frac{9}{16}$	$\frac{3}{8}$	3	10
$\frac{3}{8}$	$\frac{5}{32}$	2	18
$\frac{5}{32}$	—	1	42

(b) Dryer capacity:

From Figure 11-13 we see that dryer capacity at 7% moisture = 121 tph.

(c) Plant capacity (plant output):

$$\text{Plant capacity (tph)} = \frac{\text{dryer capacity (tph)} \times 10^4}{(100 - \% \text{ bitumen}) \times (100 - \% \text{ fines})} \qquad (11\text{-}3)$$

$$= \frac{121 \times 10^4}{(100 - 6) \times (100 - 5)} = 135.5 \text{ ton/hr}$$

(d) Determine feeding rate.

Determine the portion of the total mix supplied by each mix component (i.e., bins #1 to #4, mineral filler (fines), asphalt) and find feed rate of each.

Component	Fraction	Total (tph)	Feed Rate (tph)
Asphalt	0.06	135.5	8.1
Aggregate made up of:	0.94	135.5	127.4
Bin #1	0.42	127.4	53.5
Bin #2	0.18	127.4	22.9
Bin #3	0.10	127.4	12.7
Bin #4	0.25	127.4	31.9
Fines	0.05	127.4	6.4
Total aggregate	1.00	127.4	127.4
Total	1.00	135.5	135.5

(e) Gradation control unit feeder speeds, feeder gate openings, fines feeder calibration, and asphalt metering pump calibration procedures are determined by individual plant characteristics.

(f) Determine initial cold feed gate settings.

Figure 11-12 indicates the percentage of each type of aggregate used in the blend. The amount of each of the components that must be supplied per hour is found by multiplying the total weight of aggregate per hour by the component fraction as follows:

Aggregate	Fraction	Total (tph)	Feed Rate (tph)
A	0.45	127.4	57.3
B	0.30	127.4	38.2
Sand	0.20	127.4	25.5
Fines	0.05	127.4	6.4
Total			127.4

Cold bin assignment and method of calibrating the cold feed gates will depend on the particular plant design.

Quality Control

Proper quality control of plant mixed bituminous requires frequent sampling and testing. Proper sampling and testing procedures are described in Reference 5 at the end of this chapter. Table 11-2 lists a number of possible causes for deficiencies in the plant product. Care must also be taken during loading and hauling of the mix to insure that the mix is not degraded during these operations. Trucks must be clean and free of moisture before loading. Insulated or covered trucks may be required to insure that the mix arrives at the job site at a temperature within the specified range.

11-5 SUPPORT AND MAINTENANCE EQUIPMENT

In addition to the major pieces of equipment previously described, there are a number of items of equipment used to support and maintain bituminous construction. Equipment used for hauling bituminous products include both railroad tank cars and tank trucks. Railway tank cars range from about 6,500 to 10,000 gal in capacity. Tank cars and trucks are usually equipped with heating coils and may be insulated. Tank trucks frequently range from 2,000 to 5,000 gal in capacity.

Asphalt heaters are used to supply heat to storage tanks, transport vehicles, and asphalt melters. Both steam and hot oil heaters are commonly used. Asphalt melters are designed to melt asphalt from steel drums and heat the asphalt to pumping temperatures. The asphalt kettle may also be used to dedrum asphalt,

TABLE 11-2 *Possible causes of deficiencies in plant-mix mixtures. (Courtesy The Asphalt Institute.)*

Aggregates Too Wet	Inadequate Bunker Separation	Aggregate Feed Gates Not Properly Set	Over-Rated Dryer Capacity	Dryer Set Too Steep	Improper Dryer Operation	Temp. Indicator Out of Adjustment	Aggregate Temperature Too High	Worn Out Screens	Faulty Screen Operation	Bin Overflows Not Functioning	Leaky Bins	Segregation of Aggregates in Bins	Carryover in Bins Due to Overloading Screens	Aggregate Scales Out of Adjustment	Improper Weighing	Feed of Mineral Filler Not Uniform	Insufficient Aggregates in Hot Bins	Improper Weighing Sequence	Insufficient Asphalt	Too Much Asphalt	Faulty Distribution of Asphalt to Aggregates	Asphalt Scales Out of Adjustment	Asphalt Meter Out of Adjustment	Undersize or Oversize Batch	Mixing Time Not Proper	Improperly Set or Worn Paddles	Faulty Dump Gate	Asphalt and Aggregate Feed Not Synchronized	Occasional Dust Shakedown in Bins	Irregular Plant Operation	Faulty Sampling	Types of Deficiencies That May Be Encountered in Producing Plant-Mix Paving Mixtures.
		A												B	B				A	A	B	C	B	B	A			C			A	Asphalt Content Does Not Check Job Mix Formula
	A	A						A	A	A	A	A	A	B	B	A	A						B		A	A		C	A		A	Aggregate Gradation Does Not Check Job Mix Formula
	A	A							A	A	A	A	A	B	B	A	A						B	B				C	A		A	Excessive Fines in Mix
A			A	A	A	A	A																							A		Uniform Temperatures Difficult to Maintain
														B	B									B								Truck Weights Do Not Check Batch Weights
														B	B				A	A	B	C	B		A			C				Free Asphalt on Mix in Truck
																		B									B					Free Dust on Mix in Truck
A			A	A	A	A													A	A	B	C	B	B	A			C			A	Large Aggregates Uncoated
									A	A	A	A	A	B	B	A	A	B		A	B	C		B	A	B	C	A	A		A	Mixture in Truck Not Uniform
																		B		A				B	B	A					A	Mixture in Truck Fat on One Side
							A													A	A	B	C	B				C			A	Mixture Flattens in Truck
	A				A	A	A																								A	Mixture Burned
A			A	A	A	A													A		B	C	B					C			A	Mixture Too Brown or Gray
							B							B	A	A				A	A	B	C	B				C			A	Mixture Too Fat
					A	A	A																							A		Mixture Smokes in Truck
				A	A	A	A																							A		Mixture Steams in Truck
						A	A	A											A											A	A	Mixture Appears Dull in Truck

A—Applies to Batch and Continuous Type Plants.   B—Applies to Batch Plants only.   C—Applies to Continuous Plants.

although it is more commonly used for maintenance and repair of bituminous surfaces. Kettles range in size from 75 to 750 gal or more.

PROBLEMS

1. What is the distributor pump output required to obtain an application rate of 0.50 gal/sq yd when the spray bar length is 14 ft and the desired application speed is 3 mph?

2. Find the application rate actually obtained by an asphalt distributor under the following conditions: tank reading before spread = 1,060 gal; tank reading after spread = 480 gal; temperature correction factor = 0.95 (multiply actual volume by correction factor to obtain standard volume); width of spread = 12 ft; length of spread = 1,050 ft.

3. List in sequence the steps involved in placing a single surface treatment and identify the items of equipment commonly used for each step.

4. What is the recommended sequence for rolling a hot-mix asphalt pavement?

5. Using the dryer capacity curve of Figure 11-13 and an aggregate moisture content of 5%, determine the maximum plant output of a mix containing 8% asphalt cement and 4% mineral filler.

6. A gradation control unit has three bins, each of which holds 10 tons. Using the aggregate gradation curve of Figure 11-12 and the screen data of Figure 11-13, select the

screen sizes to be used in the gradation control unit. Tabulate the division of aggregate by these screens.

7. Calculate the feed rate (tons per hour) into the mixer of all mix components under the following conditions: aggregate gradation = Figure 11-12; screen data = Figure 11-13; bitumen content = 5%; gradation control unit = 4 bins holding 7 tons each; aggregate moisture content = 5%.

REFERENCES

1. *Asphalt Compaction Handbook.* Portland, Oregon: Hyster Company, 1966.

2. *Asphalt Handbook, The.* College Park, Maryland: The Asphalt Institute, 1970.

3. *Asphalt Mixed-in-place (Road-mix) Manual* (MS-14). College Park, Maryland: The Asphalt Institute, 1965.

4. *Asphalt Paving Manual* (S-8). College Park, Maryland: The Asphalt Institute, 1965.

5. *Asphalt Plant Manual* (MS-3). College Park, Maryland: The Asphalt Institute, 1967.

6. *Asphalt Surface Treatments and Asphalt Penetration Macadam,* College Park, Maryland: The Asphalt Institute, 1969.

7. *Bituminous Construction Handbook,* Aurora, Illinois: Barber-Greene Company, 1963.

8. *TM5-331D Construction Equipment Operations: Concrete and Asphalt Equipment.* Washington, D.C.: U.S. Department of the Army, 1969.

12

EQUIPMENT ECONOMICS

12-1 OBJECTIVES OF ECONOMIC ANALYSES

Economic analysis of construction equipment is primarily concerned with the determination of owning and operating costs for an item of equipment and the identification of the optimum economic life for an item of equipment.

Owning and operating costs (often referred to as *O & O costs*) are usually computed on an hourly basis. The cost per unit of production can then be determined by dividing the equipment's hourly owning and operating cost by its hourly production rate. An estimated cost per unit of production is necessary for bidding purposes, but actual production costs must be known for job cost control and management. In general, the objective when planning a job and selecting equipment is to minimize the cost per unit of production. However, it should be recognized that the ultimate objective of a profit-making construction organization should be to maximize profits and not necessarily to minimize costs. Although these two objectives can frequently be attained at the same time, this is not always the case.

Economic analyses supporting a decision on equipment replacement are aimed at determining the equipment replacement interval that will yield the maximum profit on the equipment investment. As will be seen shortly, the period of equipment ownership that yields the maximum profit on the equipment investment may be considerably shorter than the economic life of the equipment.

213

12-2 OWNING COSTS

General

Equipment owning costs, as the term implies, represent the cost of equipment ownership. Although these costs are usually prorated on an hourly basis for estimating and accounting purposes, they represent costs that would be incurred whether the equipment is actually used or not. Items to be included in ownership costs include the following:

1. Depreciation

2. Investment cost

3. Taxes

4. Insurance

5. Storage and miscellaneous

The composition and methods for calculating these costs are discussed in the following paragraphs.

Depreciation

Depreciation is defined as the decrease in value of property through wear, deterioration, or obsolescence. In the area of equipment economics, the calculation of equipment depreciation serves three principal purposes: (1) determining that component of owning and operating cost contributed by the equipment's decline in market value during the period, (2) determining the depreciation amount to be used in the replacement decision analysis, and (3) evaluating tax liability. It should be noted that it is possible (and legal) to use different depreciation schedules for tax purposes than are used for the other purposes. However, current tax rules of the Internal Revenue Service (IRS) provide that any cash or trade-in allowance above the equipment's depreciated (book) value received when the equipment is sold or traded will be treated as income. Thus, there is no long-term tax advantage to be gained by depreciating the item too rapidly. However, an understatement of depreciation has the effect of overstating the owner's profit for that period and results in an unnecessary tax burden. Therefore, every effort should be made to depreciate the equipment as realistically as possible. It is suggested that a record of auction values for similar equipment be used to establish the equipment's salvage value and to assist in the selection of the depreciation method to be used.

The amount to be depreciated is based on the initial cost of the equipment less the value of its tires. Because tires are a high wear item, their cost will be included in operating costs. Since initial cost is used for calculating all of the elements of owning cost, it is important that initial cost be accurately stated. Initial cost should represent the total delivered cost of the equipment including transportation, blocking or packing charges, unloading, initial assembly and servicing, sales taxes, and import duties when applicable. The amount to be depreciated may or may not include the salvage value, depending on the method of depreciation used as explained below.

Salvage value should correspond as closely as possible to the return that the item will yield at the end of the depreciation period. This may be based on historical data or used equipment auction prices. However, Internal Revenue Service regulations establish a minimum salvage value for depreciation; currently it is 10% of the initial cost.

The equipment life to be used for depreciation purposes should represent the equipment's expected economic or useful life. This will depend on the type of equipment involved, the conditions under which it will operate, and the practices in the industry. The Internal Revenue Service has established useful life guidelines for a number of classes of equipment. (The useful life for general construction equipment is currently 5 years.) The use of IRS guideline lives will simplify the calculation of depreciation for tax purposes. However, the Internal Revenue Service will accept longer or shorter periods if it can be demonstrated that these are reasonable and appropriate. Internal Revenue Service rules specify the procedures used for determining the reasonableness of depreciation procedures.

There are a number of methods for calculating depreciation. However, three methods are most commonly used. These are the straight line method, the sum-of-the-years-digits method, and the double declining balance method.

Procedures for calculating depreciation by each of these methods are given below. The sum-of-the-years-digits and double declining balance methods are often referred to as *accelerated depreciation methods* because they provide a larger depreciation in the early years of equipment life than does the straight line method.

It should be pointed out that the cost of equipment ownership may also be calculated by the methods of engineering economics. Readers interested in obtaining information on the principles of engineering economics should consult one of the many references available on the subject.

Straight Line Method

In the straight line method the amount to be depreciated is spread uniformly over the expected life of the equipment. The amount to be depreciated represents the initial cost less salvage value and less tires when applicable.

$$D_{\text{annual}} = \frac{\text{cost} - \text{salvage} \, (-\text{tires})}{\text{life (years)}} \tag{12-1}$$

Example 12-1

PROBLEM: Find the annual depreciation and book value at the end of each year for a crawler tractor having an initial cost of $35,000, salvage value of $5,000, and an expected life of 5 years using the straight line method.

Solution:

$$D = \frac{35,000 - 5,000}{5} = \$6,000/\text{yr.}$$

Year	Depreciation	Book Value (end of period)
0	0	35,000
1	6,000	29,000
2	6,000	23,000
3	6,000	17,000
4	6,000	11,000
5	6,000	5,000

Sum-of-the-Years-Digits-Method

In the sum-of-the-years-digits method the amount to be depreciated is found in the same manner as for the straight line method. However, the amount of depreciation for each year varies. The depreciation for a particular year is found by multiplying the amount to be depreciated by a depreciation factor (Equation 12-2). The denominator for the depreciation factor is the sum of the years' digits for the depreciation period; i.e., $1 + 2 + 3 + 4 + 5 = 15$ for a 5-year life. The numerator of the factor is the particular year's digit taken in *inverse* order; for the first year of a 5-year life, use 5, for the second year, use 4, etc. Thus, for the first year of life, the depreciation factor will be $\frac{5}{15}$, for the second year, $\frac{4}{15}$, etc. The procedure is illustrated in Example 12-2.

$$D_n = \frac{\text{year digit}}{\text{sum of digits}} \times \text{amount to be depreciated} \qquad (12\text{-}2)$$

Example 12-2

PROBLEM: Find the annual depreciation and book value at the end of each year for the tractor of Example 12-1 by using the sum-of-the-years-digits method.

Solution:

Find the depreciation for each year by using Equation 12-2.

$D_1 = \frac{5}{15} \times (35,000 - 5,000) = \$10,000.$

$D_2 = \frac{4}{15} \times 30,000 = 8,000.$

$D_3 = \frac{3}{15} \times 30,000 = 6,000.$

$D_4 = \frac{2}{15} \times 30,000 = 4,000.$

$D_5 = \frac{1}{15} \times 30,000 = 2,000.$

Year	Depreciation	Book Value (end of period)
0	0	35,000
1	10,000	25,000
2	8,000	17,000
3	6,000	11,000
4	4,000	7,000
5	2,000	5,000

Double Declining Balance Method

In using the double declining balance method the annual depreciation factor is found by dividing 200% by the equipment life in years. The amount of depreciation for a particular year is then found by multiplying the equipment's book value at the beginning of the year by the depreciation factor (Equation 12-3). Note that the salvage value does not enter into Equation 12-3. However, the value of the equipment may never be reduced below its salvage value. Therefore, in the later years of equipment life the amount of depreciation may have to be reduced to insure that book value does not go below salvage value. This is illustrated in Example 12-3.

$$D_n = \frac{200}{N} \times \text{book value @ beginning of year} \qquad (12\text{-}3)$$

Example 12-3

PROBLEM: Find the annual depreciation and book value at the end of each year for the tractor of Example 12-1 by using the double declining balance method.

Solution:

Depreciation rate $= \dfrac{200}{5} = 40\%$ per year.

$D_1 = 0.40 \times 35,000 = 14,000.$

$D_2 = 0.40 \times (35,000 - 14,000) = 8,400.$

$D_3 = 0.40 \times (21,000 - 8,400) = 5,040.$

$D_4 = 0.40 \times (12,600 - 5,040) = 3,024;$ use $2,560.†

$D_5 = 0.†$

Year	Depreciation	Book Value (end of period)
0	0	35,000
1	14,000	21,000
2	8,400	12,600
3	5,040	7,560
4	2,560	5,000
5	0	5,000

Investment, Tax, and Insurance

For the purpose of estimating owning costs, the cost of interest, taxes, and insurance are usually combined. A total percentage rate representing investment

---

†Note: Since $3,024 would reduce the book value to less than $5,000, only ($7,560 − $5,000 =) $2,560 may be taken as depreciation for the fourth year. Depreciation for the last year will then be zero.

cost, taxes, and insurance is multiplied by the equipment's value to obtain the annual investment, tax, and insurance cost. Although some authorities recommend the use of an average investment over the life of the equipment as the basis for determining such costs, the use of an average investment during a particular year gives a more realistic result. In this method the funds that have been allocated for depreciation are available for investment and are allowed to earn the company's current rate of return. Interest, tax, and insurance cost is then based only on the equipment's average book value during the year. Such a system reflects more accurately the cost of these items during a particular year.

Storage and Miscellaneous

These items represent the cost of storage space, facilities, and labor used to protect the equipment when it is not on a job. It should include rent and maintenance for storage yards and buildings, the wages of watchmen, the expense of handling the equipment in and out of storage, and associated direct overhead (not general overhead). These costs can be calculated on an annual basis and prorated to the equipment. If calculated on a percentage of equipment value basis, this percentage can be added to the interest, tax, and insurance percentage to obtain a combined cost percentage.

Total Owning Cost

The total cost of equipment ownership is found by summing the above categories of costs. These costs are usually computed on an annual basis and then reduced to an hourly cost by using the estimated number of operating hours during the year.

Investment Credit

In order to encourage industry to modernize production facilities, Federal income tax regulations at times provide a tax credit for the purchase of equipment. A typical investment credit used during recent years allowed a tax credit equal to 10% of the investment for equipment having a useful life of 7 years or more, with a lesser allowance for equipment having a shorter life. Investment credit amounts to a direct reduction in tax for the year of purchase. It may be treated as a reduction in the cost basis of the new piece of equipment. However, the usual method of handling investment credit does not change the cost basis and resulting depreciation. Rather, it treats the investment credit as a reduction in owning cost for the year of purchase. Hence, this credit would be subtracted from the total owning costs prior to converting costs to an hourly basis. Current IRS tax regulations should be studied for up-to-date information on investment credit procedures.

12-3 OPERATING COSTS

Elements of Cost

Operating costs include all costs directly associated with the operation of the equipment and thus vary with the amount and conditions of usage. Operator's wages may be considered a part of operating costs, but they are usually treated as a separate item which is added at the end to obtain total hourly owning and operating costs.

The principal elements of operating cost are

1. Gasoline for starting

2. Lubricants and hydraulic fluids

3. Filters

4. Periodic service and minor adjustments

5. Fuel

6. Repairs

7. Tires

Methods of estimating each of these items are discussed below.

Service Costs

Manufacturers provide consumption data or average costs for gasoline, lubricants, hydraulic oil, and filters for their equipment under average operating conditions. When consumption data are used, hourly consumption adjusted for operating conditions is multiplied by cost per item to find hourly cost of these consumables. Labor costs for service are then estimated based on the planned maintenance program and prevailing wage rates.

A quick estimate of hourly service costs for rubber-tired equipment can be made by using the TEREX procedure of multiplying hourly fuel cost by a service factor (see Reference 7 at the end of this chapter). Service factors suggested are $\frac{1}{5}$ for favorable conditions, $\frac{1}{3}$ for average conditions, and $\frac{1}{2}$ for severe conditions. Thus, the hourly service cost of a wheel tractor under average conditions would be estimated at one-third of the hourly fuel cost.

Fuel Cost

The hourly cost of fuel is found by multiplying the fuel consumption in gallons per hour by the cost of each gallon of fuel. The most accurate method for determining hourly fuel consumption is by actual measurement under similar job conditions.

However, when estimates are required, full load fuel consumption can be closely approximated by using Equation 12-4 or by using the equipment manufacturer's consumption data.

$$\text{full load fuel consumption} = 0.06 \times \text{horsepower} \qquad (12\text{-}4)$$
$$\text{(gal/hr)}$$

Since equipment rarely works continuously at full load, it is necessary to convert full load fuel consumption to consumption under the expected conditions. This is accomplished by multiplying full load consumption by a use or load factor. The manufacturer's guide or Table 12-1 may be used for this purpose. However, when possible, historical data should be used.

TABLE 12-1   *Fuel consumption factors ($\%$ of full load)*

Type Equipment	Low	Load Conditions Average	High
Clamshell and dragline	40	50	60
Cranes	30	40	50
Graders	45	60	85
Loader, track	50	75	90
Loader, wheel	45	60	85
Off-highway truck	25	35	50
Scraper, elevating	50	65	80
Scraper, standard	45	60	75
Scraper, tandem	45	65	80
Shovel and hoe	50	60	70
Tractor, crawler	45	60	80
Tractor, wheel	50	65	85
Wagons	50	65	80

Repairs

This item includes the cost of all maintenance and repair except for routine service and the replacement of high wear items (such as ripper tips and shanks and grader cutting edges). Repair cost constitutes the largest single item of operating expense for most construction equipment. Repair costs depend on equipment application, operating conditions, and maintenance standards. As would be expected, average repair costs are relatively low for new machines and rise as the equipment ages.

Manufacturers often suggest the use of a repair factor multiplied by hourly depreciation cost or initial cost less tires to obtain average hourly repair cost. Although this method is valid for average repair cost over the life of a machine, it can be very inaccurate when used for estimating repair cost in the early or late years of equipment life. This method should not be used with accelerated depreciation methods because it yields a repair cost versus time curve whose shape is exactly the reverse of the true curve.

The use of Equation 12-5 is suggested as a more accurate method for estimating repair cost during any particular year of equipment life.

$$\text{year's repair cost} = \frac{\text{year digit}}{\text{sum of digits}} \times \text{total repair cost} \qquad (12\text{-}5)$$

The method is similar to that used for finding depreciation by the sum-of-the-years-digits method except that the year digits are used in their normal order; i.e., 1 for the first year, 2 for the second year, etc. Although a similar technique using declining balance procedures may be used, a study of the repair cost histories of a large number of machines indicates that the sum-of-the-years-digits method yields results which most closely correspond to actual costs. In order to use this method, it is necessary to estimate total lifetime repair costs. This can be done by using historical data or by multiplying initial cost less tires by the appropriate factor in Table 12-2. The use of this procedure for estimating repair cost is illustrated in Example 12-4.

TABLE 12-2   *Typical lifetime repair costs (% of initial cost less tires)*

Type Equipment	Operating Conditions Favorable	Average	Unfavorable
Crawler tractors	85	90	95
Graders	45	50	55
Off-highway trucks	70	80	90
Scrapers	85	90	105
Track loaders	85	90	105
Wagons	45	50	55
Wheel loaders	50	60	75
Wheel tractors	50	60	75

Example 12-4

PROBLEM: Estimate the annual and hourly repair cost for the second year of operation of a crawler tractor costing $40,000 and having a 5-year life. Operating conditions are average and the machine will operate 2,000 hr during the year.

Solution:

Total repair cost factor = 0.90 (Table 12-2).

Lifetime repair cost = $40,000 × 0.90 = $36,000.

Repair cost for second year of operation:
Year's cost = $\frac{2}{15}$ × $36,000 = $4,800 (Equation 12-5).
Hourly repair cost = $\frac{\$4,800}{\$2,000}$ = $2.40.

Tire Cost

The cost of tire repair and replacement is another major item of expense for rubber-tired equipment. Among the components of operating cost, tire cost is usually exceeded only by general repair cost. Tire cost is also among the most difficult to estimate because of the difficulty in accurately estimating tire life.

Estimation of tire life can best be made by using good tire records. If these are not available, a rough estimate can be obtained by the use of Table 12-3. Tire replacement cost is divided by tire life in hours to find hourly tire replacement cost. Additional considerations in selecting tires and estimating tire life are contained in Chapter 13.

TABLE 12-3   *Typical tire life (hr)*

Equipment	Favorable	Operating Conditions Average	Unfavorable
Dozers and loaders	3,200	2,100	1,300
Motor graders	5,000	3,200	1,900
Scrapers:			
Single-engine	4,600	3,300	2,500
Tandem-powered	4,000	3,000	2,300
Push-pull and elevating	3,600	2,700	2,100
Trucks and wagons	3,500	2,100	1,100

Tire repair cost is usually estimated as a percentage of tire replacement cost. Unless good historical data are available for estimating tire repair costs, an estimate of 13 to 17% of tire replacement cost is suggested.

Special Items

As mentioned earlier, the cost of replacing special high wear items such as ripper tips, shanks and shank protectors, and grader cutting edges should be included in operating costs. The cost of any unusual items should also be included here. Such costs are then converted to an hourly basis.

Operators' Wages

After owning and operating costs have been determined, the operator's hourly wage is added to determine total hourly owning/operating cost for the equipment. Be sure to include workman's compensation, social security, fringe benefits, and overtime or premium pay when calculating operators' hourly wages.

12-4 TOTAL OWNING AND OPERATING COST

A suggested summary sheet for estimating equipment owning and operating cost is shown in Figure 12-1. Remember that these costs represent direct costs only and do not include overhead. The calculation of total owning and operating costs for a wheel tractor scraper is illustrated in Example 12-5 below.

Equipment Identification _____

OWNING COST:

Depreciation	$_____/hr	
Interest, tax and insurance	$_____/hr	
Storage and miscellaneous	$_____/hr	
	TOTAL	$_____/hr

OPERATING COST

Fuel	$_____/hr	
Service	$_____/hr	
Repairs	$_____/hr	
Tires	$_____/hr	
Special Items	$_____/hr	
	TOTAL	$_____/hr

OPERATOR'S WAGE	$_____/hr
TOTAL O & O COST	$_____/hr

Figure 12-1 Suggested owning and operating cost summary form.

Example 12-5

PROBLEM: Find the expected hourly owning and operating cost for the first year of operation of the wheel tractor scraper described below:

Cost delivered = $115,000.

Tire cost = $15,000.

Estimated life = 5 years (2,000 hr/year).

Salvage value = $10,000.

Depreciation method = sum-of-the-years-digits.

Interest rate = 10%.

Tax, insurance, and storage = 8%.

Load conditions = average.

Tire life = 3,000 hr.

Rated power = 415 hp.

Fuel price = $0.25/gal.

Operator's wages = $6.00/hr.

Solution:

OWNING COST:

Depreciation cost:

$$D_1 = \frac{5}{15} \times \frac{(115,000 - 15,000 - 10,000)}{2,000} = \$15.00/\text{hr.}$$

Interest, tax, insurance, and storage:

Investment @ start of year = \$115,000.
Investment @ end of year = \$115,000 − 30,000 = \$85,000.
Average investment = $\dfrac{115,000 + 85,000}{2}$ = \$100,000.

Cost rate = interest + tax, insurance, and storage = 10 + 8 = 18%.

Interest, tax, insurance, and storage = $\dfrac{\$100,000 \times 0.18}{2,000}$ = \$9.00/hr.

Total owning cost = 15.00 + 9.00 = \$24.00/hr.

OPERATING COST:

Fuel cost:

Expected consumption (Equation 12-4 and Table 12-1).
Consumption = 0.06 × 415 × 0.60 = 14.9 gal/hr.
Fuel cost = 14.9 × \$0.25 = \$3.73/hr.

Service cost:

Estimate $\frac{1}{3}$ of fuel cost.
Service cost = $\frac{1}{3}$ × \$3.73 = \$1.24/hr.

Repair cost (Equation 12-5 and Table 12-2):

Lifetime cost = (\$115,000 − \$15,000) × 0.90 = \$90,000.
Repair cost = $\dfrac{1}{15} \times \dfrac{\$90,000}{2,000}$ = \$3.00/hr.

Tire cost:

Replacement = $\dfrac{\$15,000}{3,000}$ = \$5.00/hr.
Repair (average) = \$5.00 × 0.15 = \$0.75/hr.
Total = \$5.00 + \$0.75 = \$5.75/hr.

Special items: None.

Total operating cost = 3.73 + 1.24 + 3.00 + 5.75 = \$13.72/hr.

Operator's wages: \$6.00/hr.

Total O & O cost = 24.00 + 13.72 + 6.00 = \$43.72/hr.

12-5 THE REPLACEMENT DECISION

Introduction

Selection of the optimum time at which to replace a piece of construction equipment may greatly increase the profit that the machine returns to its owner. In spite of this, many owners replace equipment rather haphazardly. Typically, these owners replace equipment when it requires major repairs or an overhaul, when extra fund are available for equipment purchase, or when preparing to start a new project. An analysis of the economics of equipment replacement using the methods discussed below will enable the owner to make replacement decisions that maximize the profits returned by the equipment.

Factors to Be Considered

In the method of replacement analysis described below only costs which vary significantly with the age of the machine are considered. We will consider five typically important factors. These are

1. Depreciation and replacement cost

2. Investment cost

3. Repair cost

4. Downtime cost

5. Obsolescence cost

Other items, such as fuel costs, service costs, tire costs, etc., may be included if they are significant in a particular case. The other factors making up owning and operating costs which are not related to the age of the equipment should be omitted from the replacement analysis.

Cumulative Cost Method

This method of analysis sums up the cost in the above categories for each year of equipment life and divides them by the cumulative operating hours accrued at the end of each year. The optimum time for replacement is then identified as that period which yields the lowest cumulative cost per hour of operation. Procedures for calculating yearly costs of each of the five replacement factors are described below.

Depreciation and Replacement Costs

The basis and methods for calculating depreciation cost were described in Section 12-2. For the purpose of this analysis, it is important that the depreciation assigned to each year of equipment life correspond as closely as possible to the actual depreciation in value of the machine.

An additional factor not previously considered is the cost of a comparable machine at the time of replacement. Historically, construction equipment prices have shown an average increase of about 5% per year over the past 20 years or so. This figure may be used to project replacement cost or it may be adjusted to reflect current market trends. In any case, the difference between the cost of a replacement machine and the trade-in or resale value of the present machine represents the capital outlay required at the time of replacement. This sum also represents the depreciation and replacement cost of the equipment at that time.

Investment Costs

The method of computing costs related to the capital investment (interest, tax, insurance, and storage) for each year was described in Section 12-2. The same method should be used in calculating these costs for the replacement analysis. As you recall, the average value of the equipment during the year is used as the basis for computing the investment cost during any year.

Repair Costs

The method of apportioning total repair costs over the life of the equipment was also covered previously. In reality, costs are incurred on an irregular basis. However, the method used previously evens out these irregularities to yield a smooth cost versus time curve. Since many of the periodic costs may be defined in terms of equipment life, it is possible to separate out such costs. The cost of engine overhauls, transmission rebuilds, etc, can then be applied to the appropriate year of equipment life. The total repair cost for a specific year would then be the sum of scheduled major repairs and an apportioned cost for all unscheduled repairs. In assigning yearly repair costs for the replacement analysis either the method of Section 12-3 or the modified method just described may be used.

Downtime Costs

In calculating owning and operating costs, consideration was not given to the effect of equipment breakdown. In reality, equipment availability decreases with equipment age and this loss of availability does have a cost implication. In order to compensate for equipment downtime, it is necessary to have extra equipment available at the job site or on call to maintain the planned production rate. One method of assigning downtime costs to a particular year of equipment life is to use the product of the estimated percentage of downtime (percent nonavailable) multiplied by the planned hours of operation for the year multiplied by the hourly cost of a replacement or rental machine.

The availability of equipment during each year of life will vary considerably depending on the make and model of equipment, the conditions of use, and the standards of maintenance. An analysis of the owner's historial data should yield the best estimate of equipment availability. If such data are not available, the use of the equipment manufacturer's data is suggested.

Obsolescence Costs

Another factor not considered in calculating owning and operating cost which influences the optimum economic life of equipment is obsolescence or the increasing productivity of new machines. As was the case with repair cost, the cost associated with obsolescence may not occur evenly but rather tends to rise sharply upon the introduction of a new model of equipment. Nevertheless, during recent years the increase in the productivity of construction equipment has averaged about 5% per year.

The cost of obsolescence for a particular year is found in a manner similar to that used for downtime cost. That is, the percentage loss in production is multiplied by the planned hours of operation for the year multiplied by the hourly cost of a replacement machine.

Optimum Equipment Life

Using the cumulative cost method, we find the cumulative cost of all factors at the end of each year of life. The result is then divided by the cumulative number of hours of operation to yield a cumulative cost per cumulative hour of operation at the end of each year. The equipment life corresponding to the lowest cost per cumulative hour of operation is then selected as the optimum equipment life for replacement. The method is illustrated in Example 12-6.

Example 12-6

PROBLEM: Using the cumulative cost method, determine the optimum time for replacement of a crawler tractor. The tractor costs $50,000 new and will have an estimated salvage value of $5,000 at the end of 5 years. Total repair cost for 5 years is estimated at $45,000. Investment cost is estimated at 15% per year. Equipment availability is estimated at 97% the first year decreasing by 3% per year. Replacement cost is expected to rise 5% per year. Loss of productivity is also expected to be 5% per year after the first year. The cost for a replacement machine is set at $10 per hour of operation. Use sum-of-the-years-digits method for depreciation and repair costs. Equipment use is planned for 2,000 hr per year.

Solution:

Replacement cost:

$$0.05 \times \$50,000 = \$2,500 \text{ increase/year.}$$

Depreciation cost:

$$D_1 = \tfrac{5}{15} \times (\$50,000 - \$5,000) = \$15,000, \text{ etc.}$$

Depreciation and replacement costs:

	1	*2*	*3*	*4*	*5*
			End of Year		
Replacement	$ 2,500	$ 2,500	$ 2,500	$ 2,500	$ 2,500
Depreciation	15,000	12,000	9,000	6,000	3,000
Depreciation and replacement costs	17,500	14,500	11,500	8,500	5,500
Cumulative	17,500	32,000	43,500	52,000	57,500

Investment cost:

	Year				
	1	*2*	*3*	*4*	*5*
Start of year	$50,000	$35,000	$23,000	$14,000	$ 8,000
End of year	35,000	23,000	14,000	8,000	5,000
Average, investment	42,500	29,000	18,500	11,000	6,500
Investment cost	6,375	4,350	2,775	1,650	975
Cumulative	6,375	10,725	13,500	15,150	16,125

Repair cost:

Year	*Cumulative*
$R_1 = \frac{1}{15} \times 45,000 = \$\ 3,000$	$ 3,000
$R_2 = \frac{2}{15} \times 45,000 = \$\ 6,000$	$ 9,000
$R_3 = \frac{3}{15} \times 45,000 = \$\ 9,000$	$18,000
$R_4 = \frac{4}{15} \times 45,000 = \$12,000$	$30,000
$R_5 = \frac{5}{15} \times 45,000 = \$15,000$	$45,000

Downtime cost:

Year	*Cumulative*
$Dn_1 = 0.03 \times 2,000 \times \$10 = \$\ \ 600$	$ 600
$Dn_2 = 0.06 \times 2,000 \times \$10 = \$1,200$	$1,800
$Dn_3 = 0.09 \times 2,000 \times \$10 = \$1,800$	$3,600
$Dn_4 = 0.12 \times 2,000 \times \$10 = \$2,400$	$6,000
$Dn_5 = 0.15 \times 2,000 \times \$10 = \$3,000$	$9,000

Obsolescence cost:

Year	*Cumulative*
$Ob_1 = 0$	0	0
$Ob_2 = 0.05 \times 2,000 \times \$\ 10 = \$1,000$	$ 1,000
$Ob_3 = 0.10 \times 2,000 \times \$\ 10 = \$2,000$	$ 3,000
$Ob_4 = 0.15 \times 2,000 \times \$\ 10 = \$3,000$	$ 6,000
$Ob_5 = 0.20 \times 2,000 \times \$110 = \$4,000$	$10,000

Cumulative costs:

	Year				
	1	*2*	*3*	*4*	*5*
Depreciation and replacement	$17,500	$32,000	$43,500	$52,000	$57,500
Investment	6,375	10,725	13,500	15,150	16,125
Repair	3,000	9,000	18,000	30,000	45,000
Downtime	600	1,800	3,600	6,000	9,000
Obsolescence	0	1,000	3,000	6,000	10,000
Cumulative cost	$27,475	$54,525	$81,600	$109,150	$137,625
Cumulative hours	2,000	4,000	6,000	8,000	10,000
Cost/Cumulative hours	$13.74	$13.63	$13.60	$13.64	$13.76

Thus, the optimum time to replace this tractor is at the end of its third year of use when it will have cost an average of $13.60 per hour of operation.

Use of Mathematical Models

The use of mathematical models for analyzing the economics of machine re-placement dates back to at least 1923 when J. S. Taylor presented an algebraic method for solving the problem of single machine replacement. Since that time a number of improvements have been made in both the models and methods for their solution. In 1963 Douglas presented a solution to the problem by using com-puter analysis. He has since extended his model to include the effect of additional factors, including the influence of income tax laws. (See References 4 and 5 at the end of this chapter.)

Mathematical models that use computer analysis enable one to incorporate a large number of variables in the model and to obtain a solution rapidly at a reasonable cost. Douglas' model optimizes equipment life by determining the replacement age at which the net worth of machine profits is maximized. Thus, his objective is properly the maximization of profits and not the minimization of cost. A typical graph of the net worth of profits after tax versus equipment life produced by the computer model is shown in Figure 12-2. The optimum economic life is found at the peak of the curve (point *A*) where profits are maximized. Although the equipment life that will produce a profit extends for a number of years, the net worth of profits declines after point *A*. Therefore, the equipment should be

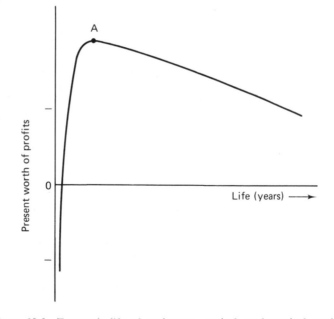

Figure 12-2 Economic life of equipment—typical mathematical model results.

replaced as close to point *A* as possible if profits are to be maximized. One of the major problems reported by Douglas in applying his model to an actual equipment fleet was the problem of determining historical cost data.

12-6 EQUIPMENT COST CONTROL

Recording Equipment Costs

Maintaining accurate equipment cost records is a must for a number of reasons. In addition to their use in making a replacement decision, cost records are required for bidding purposes, for the establishment of equipment rental rates, and for tax purposes. An analysis of cost records is also helpful in evaluating the performance of equipment by type, make, and model. The analysis will provide guidance in the procurement of replacement equipment. Unfortunately, as Douglas discovered there is a dearth of accurate equipment cost data among construction equipment owners. Thus, the purpose of this section is to point out the importance of maintaining accurate equipment cost records and to suggest some methods for accomplishing this task. Costs should be recorded by individual machine and computed on an hourly basis. Knowing the machine's hourly production makes it a simple matter to calculate equipment cost per unit of production.

Elements of Equipment Cost

In order to obtain a proper perspective on the relative value of the various elements of equipment cost, let us consider a typical construction contractor's financial analysis illustrated in Table 12-4. These figures were developed by Caterpillar Tractor Co. and were based on an actual financial analysis of a successful small earthmoving firm. Such figures will, of course, vary somewhat between different organizations. However, for most successful companies, they will vary very little from year to year in spite of changes in the total dollar value of operations. Note that consequential costs, such as downtime, were not considered, although these costs do affect job cost and production.

Records

Accumulating the cost data discussed above requires maintaining detailed cost records for each machine. These records may be in the form of log books, daily record sheets or cards, computer cards, etc. Regardless of the method used, each operator should maintain a daily record for each machine of the hours worked, the location and type of work performed, and the amount of fuel and lubricants used. Servicemen must record the time spent on servicing as well as the quantity of parts and supplies used. Repair records for the machine should be maintained on a similar basis. After these figures have been posted to a central record, it is a simple bookkeeping matter to enter and extend any necessary prices for supplies, parts, or labor.

TABLE 12-4 *Typical distribution of earthmoving company costs*

Cost Element	Percent of Total Income		
Machine costs:			66
Fixed costs:		27	
Depreciation	21		
Interest, tax, and insurance	6		
Variable costs:		39	
Operator's wages	18		
Repairs and maintenance	11		
Fuel and lubricants	8		
Equipment rental	2		
Overhead:			17
Fixed:		12	
Supervision	7		
Accounting and legal services	3		
Rent and utilities	1		
Dues and contributions	1		
Variable:		5	
Advertising	1+		
Supplies	1+		
Travel	1+		
Postage, freight, and miscellaneous	1+		
Profit			17
Total			100

(Courtesy Caterpillar Tractor Co.)

Computer methods are becoming widely used for maintaining cost data and for correlating costs to job progress. A computer program has been developed specifically to maintain tire records (see Reference 9 at the end of this chapter).

PROBLEMS

1. Find the hourly owning cost for the second year of operation of a crawler tractor under the following conditions:

    Tractor cost = $65,000.
    Salvage value = $15,000.
    Estimated life = 5 years (2,000 hr/yr).
    Depreciation method = double declining balance.
    Interest, tax, and insurance = 12%.
    Storage and miscellaneous = 4%.

2. Find the hourly owning cost for the third year of operation for the crawler tractor of Problem 1. Use the sum-of-the-years-digits method for depreciation.

3. Find the hourly operating cost for the second year of life of the tractor in Problem 1. Use the following data:

    Rated power = 300 hp.
    Fuel price = $0.45/gal.
    Load conditions = high.
    Operating conditions = unfavorable.

4. Estimate the hourly owning and operating costs for the first year of operation of a wheel tractor scraper. Use the following data:

   Delivered price = $160,000.

   Tire cost = $20,000.

   Salvage value = $40,000.

   Estimated life = 5 years (2,000 hr/yr).

   Depreciation method = sum-of-years-digits.

   Rated power = 400 hp.

   Load conditions = average.

   Operating conditions = average.

   Fuel price = $0.40/gal.

   Tire wear conditions = average.

   Interest, tax, and insurance = 12%.

   Storage and miscellaneous = 3%.

   Operator's wages = $8.00/hr.

5. Determine the optimum economic life of the scraper in Problem 4. Use cost and productivity increases of 5% per year. Downtime is estimated at 4% for the first year, increasing 4% per year.

REFERENCES

1. *Basic Estimating* (3rd ed.). Melrose Park, Illinois: Construction Equipment Division, International Harvester Company, n.d.

2. *Caterpillar Performance Handbook* (5th ed.). Peoria, Illinois: Caterpillar Tractor Co., 1975.

3. *Contractors' Equipment Manual.* Washington, D.C.: The Associated General Contractors of America, Inc., 1975.

4. DOUGLAS, JAMES, *Construction-Equipment Policy.* New York: McGraw-Hill, 1974.

5. ———, *Construction-Equipment Policy: The Economic Life of Equipment*, Technical Report No. 61. Stanford, California: The Construction Institute, Stanford University, 1966.

6. *Equipment Economics.* Peoria, Illinois: Caterpillar Tractor Co., n.d.

7. *Production and Cost Estimating of Material Movement with Earthmoving Equipment.* Hudson, Ohio: TEREX Division, General Motors Corporation, 1970.

8. *Technical Bulletin No. 2: Operating Cost Guide.* Milwaukee: PCSA Bureau, Construction Industry Manufacturers Association, 1965.

9. *Tire Records by Computer.* Peoria, Illinois: Caterpiller Tractor Co., n.d.

13

MAINTENANCE, TIRES AND SAFETY

13-1 EQUIPMENT MAINTENANCE

General

Equipment maintenance is the servicing, adjusting, and repairing of equipment. Construction equipment managers must be aware of the importance of proper equipment maintenance and the effect of equipment breakdowns on job production and costs. Some typical relationships between equipment age, downtime, and downtime costs were explored in the previous chapter. However, specific factors will vary greatly, depending on job and equipment conditions. All too frequently operators and supervisors in the field attempt to increase production by operating equipment under load conditions greater than the equipment was designed to handle. The result is premature breakdown with accompanying delays and cost increases.

Proper preventive maintenance procedures and an efficient repair system will minimize equipment failures and their consequences. Suggestions for providing effective maintenance procedures are given in the following paragraphs. Maintenance can be divided into several levels or categories. The categories that will be used here are preventive maintenance, minor repair, and major repair.

233

Preventive Maintenance

Preventive maintenance (sometimes referred to as *PM*) is routine periodic maintenance and adjustment designed to keep equipment in the best possible operating condition. It consists of a number of elements that may be compared to links in a chain. The primary links in the PM chain are the skill of the operator, the manner in which the equipment is used, proper fuel handling, proper equipment lubrication, and correct periodic adjustment. If any of the links of this chain fails, the result will be premature equipment breakdown.

Equipment manufacturers have developed specific lubrication, servicing, and adjustment procedures for each piece of their equipment. These procedures should be carefully followed. Upon request, major petroleum companies will provide guidance on the proper lubricants to be used under specific operating conditions. Because of the hostile environment in which construction equipment operates, it is important to keep dust, dirt, and water from entering the engine and other mechanical assemblies of the equipment. This requires special precautions to keep fuel clean and to keep the equipment's air and fuel filters operating properly. Some precautions to be observed in fuel handling include the following:

1. Avoid using barrels for storing and transferring fuel whenever possible. If barrels must be used, allow time for contaminants to settle and use barrel pumps for removing fuel.

2. Whenever possible, store fuel at the job site in tank trucks or storage tanks. Storage tanks should be located away from haul roads and other sources of dust. Storage tanks should be sloped or equipped with sumps and equipped with drains for removing water and sediment from the bottom of the tank.

3. Water-sensing compounds may be used to check for the presence of water in tanks. If water is detected, the tank, filters, and water separators should be drained until all traces of water are removed.

4. Dispensing equipment should be equipped with filters and water separators to remove any contaminants present. Hose nozzles and filler openings should be equipped with caps and wiped clean before the start of fuel transfer.

5. Fill equipment fuel tanks at the end of each day's operation to reduce moisture condensation in the tanks during the night. This is especially important during cold weather operations.

6. Fuel samples should be taken upon receipt of fuel from the distributor and again during fueling operations. Samples should be analyzed for cleanliness. The source of any contamination should be located and corrected.

7. Avoid refueling equipment in the open when it is raining, snowing, or very dusty.

Clean air is equally as important as clean fuel is to an engine. Air cleaners must be serviced at the interval recommended by the equipment manufacturer or more often under extremely dusty conditions. Filter housings and precleaners must be blown out, vacuumed, or wiped clean when filters are changed. Crankcase breathers should be serviced in the same manner as air filters.

Wipe fittings clean before lubricating. Crankcase oil must also be kept clean. Precautions similar to those used in fueling should be observed when replacing crankcase oil. Filters must be replaced and the filter housing wiped clean at specified intervals. When changing filters, observe the condition of the seals on filter housing caps and check for ruptured filters. If ruptured filters are found, shorten the interval for oil and filter change. After filling the crankcase, run the engine for a few minutes, check for leaks, observe oil pressure, and check dipstick level. Transmissions and hydraulic control systems should be serviced in much the same manner as the engine oil system. Use the specified hydraulic fluids and filters. The system should be checked frequently for leaks. Air entering the hydraulic control system will cause rough operation and a chattering noise.

Oil analysis programs consisting of periodic sampling and laboratory analysis of equipment lubricants are rapidly becoming common. Samples are analyzed by spectrometry and physical tests to determine the presence of metals, suspended and nonsuspended solids, water, or fuel in the lubricants. It has been found that the oil circulating in an engine reflects the condition of the engine by the presence of wear particles, contaminants, etc. Thus, oil analysis provides an excellent guide to the internal condition of an engine. The interpretation of laboratory results is not based primarily on indicator levels obtained for a single test but rather on the deviation from the equipment's historical pattern. Such programs have often significantly reduced repair and maintenance costs by allowing adjustment of maintenance intervals to fit job conditions and by detecting potential failures prior to a breakdown. The reduction in downtime cost can also be significant.

Many of the bearings on modern construction equipment are sealed to prevent entrance of dirt and water and to reduce the frequency of required lubrication. Except for permanently lubricated sealed bearings, enclosed bearings must be lubricated at the interval specified by the equipment manufacturer. Exposed mechanical parts (gears, cables, etc.), however, require different treatment. They should receive only a light covering of the specified lubricant. A heavy coating of lubricant on exposed parts collects dust and dirt and results in rapid wear of moving parts.

PM Indicators

Preventive maintenance indicators (PM indicators) are conditions which may be readily observed by an equipment manager and provide a guide to the maintenance condition of the equipment. Although a number of these indicators have been developed for specific equipment, the following are of general application:

1. After the equipment has been standing idle for several hours, check on the ground for grease, oil, or water spots that will indicate leaks.

2. Make a visual inspection of the equipment for loose bolts, leaking hoses or seals, and any unusual wear.

3. Check blades for holes or dents. Check cutting edges and end bits for excessive wear and loose bolts.

4. On crawler-type equipment check track for correct tension and loose shoe bolts. Loose track bolts are indicated if there is a shiny surface around the bolt head and if dirt is knocked off loose bolt heads by vibration. Modern equipment often uses hydraulic track adjusters which are tensioned with a grease gun. Track adjustments must be made on the job since certain soils tend to tighten the track during operation. The undercarriage should be kept as free as possible of mud and debris to prevent loss of power and unnecessary track wear.

5. Make sure that the radiator is free of debris and that the radiator core openings are clean. Fan belts and other drive belts should be in good condition and properly tensioned. Using a belt tension gauge for checking belt tension is strongly recommended.

6. Be sure that fuel, oil, hydraulic fluid, and water are at their proper levels. Gauges should be checked for condition and proper operation.

7. Insure that the air cleaner and precleaner are serviced as required.

8. Check that cables and sheaves are clean and properly lubricated and that cables are free of kinks and broken strands.

9. Be sure that the operator's floor is clear of hazards and loose objects and is free of grease and oil.

10. Check that the battery is clean and undamaged. Battery cables should not be frayed and connections should be tight.

11. Adjust brakes and clutches properly.

12. Check tires for proper inflation and check the condition of treads and sidewalls. Improper inflation is the major maintenance and safety problem.

13. Watch for dark smoke coming from the exhaust after the engine has warmed up. Smokey exhaust usually indicates a clogged air intake or fuel problem (damaged fuel injector, wrong fuel, etc).

Maintenance Organization

Routine maintenance and servicing may be performed in the open or at a covered job-site facility by either the operator or a service team. Servicing is usually performed in the open except when frequent adverse weather conditions prevail, but service areas should be located away from haul roads and other sources of dust. Both operator maintenance and crew maintenance systems have been sucessfully employed. However, specialized service teams equipped with mobile power lubrication and fueling equipment have been used most successfully by many contractors. This system allows maintenance to be performed during shift breaks, at the end of the day, or at staggered intervals during equipment operations. Minor repairs may be performed on the job by mobile repair teams or at an on-site repair facility. Equipment requiring major repairs is normally brought to an equipment dealer or company shop for repair. Large projects, particularly at an isolated location, may justify the establishment of a major repair facility near the job site.

Since the supply of repair parts is a frequent problem at all maintenance levels repair parts stockage policies must be carefully developed based on experience, manufacturer's demand data, and job conditions. Frequently the stockage and repair by replacement of assemblies and subassemblies will result in a lowering of total maintenance costs. The defective assemblies which are removed are later repaired or exchanged for rebuilt assemblies. When computers are available, they may be used advantageously to maintain data on equipment repair and cost history as well as to regulate repair parts stockage.

13-2 TIRES

General

As we saw in Chapter 12, tire cost is a major component of operating cost for rubber-tired equipment. Hence, it is important to understand the types of tires and treads that are available for construction equipment and the considerations involved in selecting the most appropriate tire for a particular application.

Tires may be classified as bias ply tires, radial tires, and special construction tires. Bias ply tires use a carcass consisting of alternating layers of cord (plies) crossing each other at about a 45°-angle to each other. Bias ply tires may or may not have circumferential cord layers (belts) running under the tread to provide protection to the carcass plies. If equipped with belts, such tires are called *belted/bias tires*. Radial tires utilize a carcass constructed of cords wrapped radically from bead to bead. Steel cords are frequently used with radial construction. Belts are always used with radial construction and frequently are made of steel cords. An example of a special construction tire design is the Caterpillar Beadless Tire developed specifically for use on construction equipment. This tire uses a sealed oval air chamber similar to a doughnut held in place by a special two-piece rim. The rim does not form a part of the air chamber. The tire tread consists of a removable belt with attached treads, such as steel shoes. Advantages claimed for this tire include less heat buildup, greater cut protection, better stability, longer tire and tread life, and easy replacement of tread.

Another term used in tire design with which you should become familiar is *aspect ratio*. The aspect ratio of a tire is the ratio of tire section height to width.

Aspect ratios for the various common tire sections are standard tire = 1.00, wide base tire = 0.85, low profile tire = 0.65, and Caterpillar Beadless Tire = 0.45.

Tire Identification Codes

The Tire and Rim Association has developed a code identification system to be used for nonhighway tires. This code system is designed to reduce the confusion caused by the wide variety of trade names and types previously used by tire manufacturers. The basic categories (which identify the type of service for which the tire is designed) used in this system are:

C = compactor service

E = earthmover service

G = grader service

L = loader and dozer service

LS = log-skidder service

ML = mining and logging service

Numbers are used with the letter symbols listed above to designate the type of treads used. Although the specific tread types vary slightly for different service categories, the basic types identified by the numbers 1 through 7 are:

1 = smooth or rib tread

2 = traction tread

3 = rock tread

4 = rock deep tread

5 = rock extra-deep tread (L5) or rock intermediate HR† (E-5)

6 = rock maximum HR†

7 = flotation

Tire Life and Performance

Some of the principal tire life factors include tire size and type, heat, condition of the travel surface, quality of vehicle maintenance, operator skill, and inflation pressure.

The buildup of heat in tires affects tire strength and tire wear. Heat buildup is determined by load, speed and length of haul, and the ambient temperature. In recognition of the importance of the relationship between heat buildup and tire safety, tire and equipment manufacturers have developed a system for assigning

†HR = heat resistant.

maximum ton-mile per hour (TMPH) ratings to tires. Equation 13-1 is used to determine the required TMPH rating for a particular job situation.

$$\text{TMPH} = \text{average tire load (tons)} \times \text{average speed (mph)} \qquad (13\text{-}1)$$

Average tire load is found by adding the empty tire load and the fully loaded tire load and dividing by two. Average speed is calculated as the total miles traveled during a work day divided by the length of the work day in hours. (The length of the work day includes lunch breaks, rest stops, etc.) The tires used on the job must have a ton-mile per hour rating equal to or higher than the job ton-mile per hour value found by using Equation 13-1. In addition to complying with ton-mile per hour limitations, the Tire and Rim Association rating of safe load at various inflation pressures must be checked to insure that the tire being used has adequate structural capacity to carry the load. For heavy loads, it may be necessary to change to a tire having a higher ply rating. Tire pressure should be checked at least once daily.

In addition to heat buildup, tire wear is influenced by the physical condition of the travel surface. Tread wear is increased by steep grades, sharp curves, rocks and other sharp objects on the surface, and by poorly drained surfaces. Unequal or grabbing brakes, worn suspension systems, tires rubbing against the chassis, etc., also increase tire wear. Operator ability, (for example, the operator's accelerating and braking technique) and operator care may either increase or decrease tire wear.

Typical tire life for various types of equipment and job conditions were given in Table 12-3. Some tire and equipment manufacturers provide a method of estimating tire life based on a number of wear factors, such as grade, speed, load wheel position, etc. (see References 1 and 8 at the end of this chapter). Another method utilizes average mileage for a specific size tire multiplied by adjustment factors for tire loading, travel speed, rock cuts, and type of tire construction. Average mileage is then divided by average speed to yield tire life in hours (see Reference 4).

Tire Selection

The objective in selecting a tire for a specific construction equipment application is to obtain the lowest cost per hour of operation consistent with obtaining the required performance. The types of tires available and methods for estimating tire life have already been given. Some additional considerations and suggestions are presented below.

Radial tires tend to run cooler than bias ply tires, provide good flotation and traction, and have a lower rolling resistance than bias ply tires. Thus, radial tires are generally preferred for high-speed hauling applications. When working in very abrasive materials, consider using rock or extra-deep tread tires. In rock areas consider using rock treads, protective tire chains, or special construction tires.

Caterpillar Tractor Co. has made the following suggestions for selection of scraper and haul unit tires (Reference 2):

1. Primary consideration should be TMPH.

2. Use lowest ply rating meeting load requirements.

3. Use largest optional size.

4. Use thickest and most cut-resistant tread consistent with TMPH.

5. Tread should have the largest practical bar to gap ratio.

6. Consider belted construction.

13-3 SAFETY

General

Safety is an important consideration in construction for a number of reasons. Principal among these are the prevention of suffering and loss of life, reduction in insurance costs (workman's compensation and hazard insurance), prevention of delays caused by accidents, and, of course, compliance with legal requirements. This last factor has taken on increased importance since the passage of the Federal Occupational Safety and Health Act (OSHA) of 1970. OSHA provides a comprehensive system of safety regulations, inspections, and record keeping. Civil penalties for OSHA violations range up to a $10,000 fine per violation. Criminal penalties run as high as a $10,000 fine and 6 months in jail for willful violation resulting in death or for the falsification of safety records.

OSHA requirements for equipment operator protection often include rollover protection, seat belts, back-up alarms, improved brake systems, and safety guards for moving parts. In addition, OSHA sets maximum noise levels for operators and workers. Limitations on operator exposure to noise have led to a need for improved equipment instrumentation to enable the operator to determine whether his machine is operating properly without depending on the sound of the equipment's operation.

Safety Programs

A carefully planned and directed safety program is required to minimize accidents and insure compliance with OSHA and other safety regulations. To be successful, a safety program must instill a sense of safety consciousness in every worker. Safety, after all, is primarily a matter of applying common sense to the prevention of accidents. Some of the major items to be considered in planning a safety program include the following:

1. Establishing a formal safety training program for all new employees and providing periodic refresher training.

2. Organizing a supervisory safety training program for all supervisors followed up by regular visits by safety personnel.

3. Providing all necessary protective clothing and equipment.

4. Making available first aid equipment and trained personnel at the job site.

5. Providing for prompt emergency evacuation of injured employees to a medical facility.

6. Continuing review and control of job hazards.

7. Establishing a procedure for maintaining safety records and reporting accidents.

8. Finally but most important, obtaining top management interest in safety. The responsibility for planning and directing the safety program must be clearly defined.

Such a program, conscientiously executed, will minimize the occurrence of accidents and their accompanying losses.

Safety Precautions

The safety manuals published by the Construction Industry Manufacturers Association (References 3, 6, 7, 11, 12, 13) contain a number of safety rules and suggestions for safe operation of specific items of construction equipment. Following is a brief list of major safety precautions for equipment operations:

1. Don't allow workers to ride on equipment unless proper seating is provided.

2. Take positive measures to insure that equipment cannot be accidentially operated while under repair.

3. Use blocking or cribbing when mechanics or workers must work under heavy loads supported by cables, jacks, or hydraulic systems.

4. Insure that equipment is parked with the brake set, blade or bowl grounded, and ignition key removed at the end of the work day.

5. Operators and mechanics mounting equipment must use the steps and hand holds provided.

6. Any guards or safety devices removed during equipment repair must be replaced promptly.

7. Use guides or signalmen when an equipment operator's visibility is limited or when there is danger to persons located near the equipment.

8. Shut off and tag electrical circuits when electrical equipment is being repaired.

9. Clearly mark high-temperature lines and pipes to prevent accidental burns. Be especially careful when using live steam.

10. Equipment containing hot or flammable fluids must be set on a firm foundation to prevent overturning. Provide fire extinguishers or other safety equipment as required.

11. Shut down burners or engines and do not allow smoking during refueling operations.

12. Empty aggregate bins or batching plants before performing repairs.

13. Slow-moving equipment operating on highways should use flashers or other signals to warn traffic of their presence.

14. Use particular care to avoid overturning when equipment is operated on side slopes.

15. Unless overhead protection is provided, require haul unit operators to stand clear of their vehicles during loading.

16. Be sure that loads are properly secured and covered if necessary and be sure that oversize loads are properly marked before hauling.

17. When workers or equipment are located in soil excavations over 4 ft deep, the sides should be shored, sheeted, and braced or sloped to the angle of soil repose to prevent possible cave-ins.

18. Wire ropes and cables must be of the proper size for their intended usage and should be inspected at least once a week for wear and damage.

19. Never push articulated equipment if the power system is not operating. The best procedure is to move it on a heavy-equipment trailer (lowboy) to the repair shop, but it may be towed for short distances if the manufacturer's precautions are observed.

PROBLEMS

1. When should equipment fuel tanks be filled? Why?

2. Name two methods for determining whether or not water is present in a fuel storage tank?

3. Name eight PM indicators that would apply to a crawler tractor.

4. What type of construction equipment tire is represented by each of the following tire codes?
   (a) G-1.
   (b) E-3.
   (c) L-2.

5. Explain how a job ton-mile per hour rating is determined for a particular situation.

6. What two conditions must be met to insure that a particular tire is safe for an intended job application?

7. What is OSHA and what are the maximum penalties provided under its regulations?

REFERENCES

1. *Basic Estimating* (3rd ed.). Melrose Park, Illinois: Construction Equipment Division, International Harvester Company, n.d.

2. *Caterpillar Performance Handbook*. Peoria, Illinois: Caterpillar Tractor Co., 1975.

3. *Crawler Tractor/Loader Safety Manual*. Milwaukee: Construction Industry Manufacturers Association, 1973.

4. "How To Extend and Estimate Tire Life," *The WABCO Cooperator*, Summer 1974, pp. 12–14.

5. *Manual of Accident Prevention in Construction*. Washington, D.C.: Associated General Contractors of America, Inc., 1971.

6. *Motor Grader Safety Manual*. Milwaukee: Construction Industry Manufacturers Association, 1971.

7. *Off-highway Truck Safety Manual*. Milwaukee: Construction Industry Manufacturers Association, 1970.

8. *Production and Cost Estimating of Material Movement with Earthmoving Equipment*. Hudson, Ohio: TEREX Division, General Motors Corporation, 1970.

9. *Road Building Equipment: RB-224A*. Chicago: American Oil Company, 1962.

10. *Safety and Health Regulations for Construction*, Federal Register Vol. 37, No. 243, Part II. Washington, D.C.: U.S. Department of Labor, 1972.

11. *Scraper Safety Manual*. Milwaukee: Construction Industry Manufacturers Association, 1971.

12. *Wheel Type Loader/Dozer Safety Manual*. Milwaukee: Construction Industry Manufacturers Association, 1970.

13. *60 Rules on Safety* (*Cranes and Excavators*). Milwaukee: Construction Industry Manufacturers Association, 1970.

14

SYSTEM DESIGN
AND SIMULATION

14-1 CONSTRUCTION EQUIPMENT SYSTEM DESIGN

Introduction

In planning a major construction equipment operation and selecting the equipment for the operation, the entire operation should be viewed as a system. This is true whether the operation involves earthmoving or the production of a product, such as a prefabricated structure or a component of a structure. As you recall from Chapter 12, the principal economic objective of a construction project should be to maximize profit or return on investment, subject, of course, to a number of constraints. These constraints may include time allowed by the contract, anticipated weather conditions, working capital available, governmental regulations, environmental considerations, equipment, labor, and material availability, and so on.

Although the use of the planning and management techniques presented in earlier chapters will yield increased efficiency in the operation of specific equipment, even greater savings may result from a careful analysis of the project from a systems viewpoint. The natural tendency in planning a job is often to attempt to utilize equipment already on hand and to select a manner of operation that has been successfully used on a previous job. This may, of course, yield satisfactory results. However, in many cases, better results may be obtained by using the planning techniques discussed in this chapter and the following one. Remember that a savings of one cent per cubic yard on a one million cubic yard job amounts to a $10,000 increase in profit. Thus, a higher initial investment for larger or more productive equipment may, in fact, produce the greater return on investment.

In analyzing a project from a systems viewpoint, one of the major difficulties involved is the large number of different combinations of equipment that often need to be considered. As a result, extensive analyses of earthmoving problems have traditionally been limited to extremely large projects (such as dams) because of the time and manpower required to perform such analyses. Fortunately, the electronic computer has made it possible to analyze large volumes of data rapidly and at a reasonable cost. The use of the computer in simulating construction equipment operations will be discussed in Section 14-2.

Earthmoving Systems

In analyzing a typical earthmoving system (even when considering only conventional earthmoving equipment), a large number of variables may need to be considered, some of which include:

SCRAPER (LOAD, HAUL and SPREAD) OPERATION

Scraper-type:
    Push-loaded scraper:
        Scraper characteristics:
            Crawler or rubber-tired tractor
            Size
            Horsepower to weight ratio
            Number of axles
            Number of engines
        Pusher:
            Size of pusher
            Number of pushers
            Crawler or wheel tractor
    Self-loaded scraper:
        Elevating
        Push-pull
        Multi-bowl, multi-engine
            For each of these, considerations of size,
            horsepower, etc., apply

TOP-LOADED HAULER OPERATION

Haul unit:
    Truck or wagon
    Size, number of axles, horsepower, etc.
Loader:
    Shovel
    Dragline

Dozer with hopper
Wheel loader
Bucket wheel excavator
Belt loader
Elevating grader

Spreading requirements

Similar considerations would be present in analyzing rock excavation and bituminous or concrete operations.

Many of the considerations involved in estimating production and selecting equipment for different applications have been discussed in the preceding chapters. Again, these comparisons may be greatly facilitated by using computer programs such as those developed by major equipment manufacturers.

When we analyze major earthmoving projects, we should not limit our analysis to the conventional earthmoving equipment listed above. References 7 and 9 describe theoretical studies and full-scale field trials conducted in Great Britain on the use of belt conveyors as the hauling element in highway construction. The nature of the material being excavated in the full-scale test (chalk) resulted in the use of dozers and rippers for excavation rather than the bucket wheel excavators assumed in the theoretical study. The fluctuation of the rate of belt loading using this method of excavation, as well as the nature of the material itself, resulted in a lower conveyor utilization rate than had been predicted. In spite of these factors, the average cost per cubic meter of material placed ran about 13% less than the cost of using conventional equipment and methods. An analysis of potential belt conveyor application in the construction of the major new highways planned in Great Britain over a 6-year period indicated that conveyors could economically be used on about one-third of the projects at an average savings of 18% on the affected projects. In addition to the economic advantage of conveyors, the full-scale trial indicated major environmental advantages because of decreased noise and dust in the construction area. It was also found that the construction operation employing conveyors was less affected by weather conditions than were conventional construction methods.

Extremely large earthmoving projects such as dams have often considered, and sometimes adopted, specialized excavation and hauling systems. For example, the Oraville Dam project used bucket wheel excavators with a 1-mile conveyor and 12-mile rail line to move some 65 million BCY of material. The Portage Mountain Dam project used dozers to feed belt loaders, which in turn fed a 16,000-foot belt conveyor, to move material to a mixing plant. At the mixing plant, material was dumped into 100 ton-capacity wagons and hauled to the placement site. Some 74 million BCY of material were moved by this system.

For excavating and transporting oil-bearing sand to an extraction plant at its Athabasca Oil Sands project in Alberta, Canada, the Sun Oil Company used a computer simulation model to evaluate a number of equipment and operating alternatives. The basic excavation equipment studied consisted of two large bucket wheel excavators. The excavated material had to be transported only a few hundred yards initially, but the haul distance would increase to some 3 miles as the excavation moved away from the processing plant. Transportation alternatives analyzed included various combinations of belt conveyors as well as 100 ton-capacity trucks. Equipment failure rates, weather conditions, storage capacity, and extraction plant performance were all considered using the GPSS computer simulation model described in the next section. Study results indicated that one main conveyor linking the pit to the extraction plant and fed by a bench conveyor and a trunk conveyor associated with each bucket wheel excavator yielded the optimum results (see Reference 1).

14-2 CONSTRUCTION SIMULATION

Simulation Models

In Chapter 4 some of the difficulties involved in estimating the output of a loader/haul unit system and in selecting the optimum number of haul units to use with a particular loader were noted. It was seen that the errors associated with the assumptions of the "conventional" method of balancing such systems could be reduced by applying the queueing theory. However, it was noted that the queueing theory method also had a number of limitations. For example, field observations indicated that the distribution of truck loading times by a shovel did not conform to the normal distribution assumed by the queueing theory. Since the distribution used in queueing theory could not be changed, production estimates had to be multiplied by an empirical correction factor to yield an improved estimate. In addition, loader/haul unit systems made up of different size trucks or of multiple loaders could not be adequately modeled using queueing theory.

Computer based simulation models, however, allow virtually unlimited ability to specify the variables of a construction equipment operation. For example, loader and haul unit time distributions, number and capacity of trucks and loaders, management decisions concerning sequence of operations, etc, may be varied as desired. The speed of computers also permits the planner to simulate an operation for hundreds or thousands of cycles in a short time and at a moderate cost. The advantages of simulation systems for testing equipment combinations and operational procedures are thus apparent.

There are two basic methods available for constructing computer simulation models of construction operations: either writing a computer program to simulate a specific operation using conventional programming languages such as FORTRAN or using special simulation languages to develop a model. A number of major equipment manufacturers and planning firms have developed programs of the first type to simulate the performance of construction equipment under certain conditions. An example of such a program developed by the Caterpillar Tractor Co. is described in Reference 6.

Since programming a simulation model with conventional programming languages is a tedious task for those who are not professional programmers, a number of simulation languages have been developed to simplify the task. There are two basic simulation languages: one for simulating processes which represent discrete events taking place in sequence and a second for simulating continuous systems. Most construction operations are processes of the first type (discrete serial events).

Several simulation languages of the discrete serial event type have been developed and have gained acceptance. Three of the better known simulation languages are the General Purpose Simulation System (GPSS), SIMSCRIPT, and Control and Simulation Language (CSL). The GPSS language, developed by the International Business Machines Corporation (IBM), is widely available and may be easily used for simulating many construction equipment operations. Therefore, only the GPSS language will be used in the discussion and examples that follow.

General Purpose Simulation System (GPSS)

GPSS may be considered a high-level computer language. That is, a single GPSS instruction may cause the computer to accomplish many specific functions or operations. The GPSS program causes the computer to automatically keep track of events occurring in the system, maintain a time clock, and perform certain data tabulations. Additional flexibility in the use of GPSS is provided by the fact that a GPSS program may be combined with a FORTRAN program to perform functions not incorporated in GPSS itself. An application of GPSS in planning an excavation and hauling operation was described briefly in the previous section (Athabasca Oil Sands project).

Since little information is available describing the application of GPSS to construction problems, it is hoped that the descriptions and examples of GPSS models which follow will be of value to those readers having little knowledge of computers as well as to those with programming experience in computer languages such as FORTRAN. For the beginner, it will illustrate the simplicity of modeling such construction equipment systems as truck/shovel fleets. For those familiar with computers, it will provide an introduction to the use of GPSS in construction and illustrate some typical system models. To obtain additional information on GPSS procedures and capabilities, the reader should consult one of the references at the end of this chapter.

Following are several important GPSS terms that should be understood:

Transactions. These represent the serial units moving in the system (e.g., trucks or scrapers). These are introduced into the system and removed from the system as required during the progress of the simulation. Each transaction may have a number of parameters assigned to represent various characteristics of the transaction (e.g., truck capacity).

Equipment Entities. These represent elements that are used by transactions and include both *facilities* and *storages*. A *facility* (e.g., shovel) can handle only one transaction at a time, while a *storage* (e.g., several shovels) can service several transactions simultaneously.

Statistical Entities. These are used to measure system behavior and include *queues* and *tables*. For each queue (representing, for example, trucks or scrapers waiting at a loader), a record is maintained of the transactions delayed in the system and the length of delay. The GPSS program computes the average number of transactions delayed and the average length of delay for each queue. Tables may be used to collect and print out any statistic desired by the programmer.

Operational Entities. These provide the logic of the systems and control the movement of transactions.

The first step in writing a GPSS program, as in writing any computer program, is to diagram the operation of the program. IBM has developed special block symbols to represent GPSS instructions. The use of these symbols to draw the block diagram will facilitate writing the program instructions. However, those readers familiar with conventional computer flow diagrams may prefer to continue using conventional flow diagrams, at least initially. Example 14-1 illustrates the use of both types of diagrams in constructing a system model.

The card punch input format for a GPSS program provides for three distinct data fields on each card. These are the Location Field (columns 2–6), the Operational Field (columns 8–18), and the Operand Field (columns 19–72). Columns 1, 7, and 73–80 are left blank. An asterisk appearing in column 1 of a card will cause the card to be treated as a COMMENT card and will not affect program operation. The use of each of these three fields is described below.

Location Field. This is used when the programmer desires to assign a label to a block. A label may be a number or any combination of alphabetical and numerical characters.

Operation Field. This contains a word or abbreviation that identifies the block type or the operation to be performed.

Operand Field. This field is subdivided into a variable field and a comments field. The variable field provides for as many as seven subfields (designated A to G) which are separated by commas. The number of subfields used on a particular card depends on the type of card and the desires of the programmer. It is not necessary to designate subfields if they are not required. However, if lower-ordered subfields are omitted prior to using other subfields, then separating commas must be inserted to indicate the omission. For example: if subfields A and B are omitted but subfield C is used, the entry in the operand field would read ", , C." The end of the variable field is indicated by the first blank character encountered after column 19. Thus, any information encountered after a blank in the variable field is treated as a comment.

Several types of cards and the meanings of other GPSS terms will be explained as they are encountered in the following examples. Although almost any statistic desired by the programmer can be maintained and provided as output, the following statistics are normally generated and printed out:

Storage Block. Capacity of the storage, the average contents, average utilization, number of entries, average time a transaction used the storage, the current contents, and the maximum contents.

Facility Block. Same information as storage.

Queue Block. Maximum contents, average contents, total entries into the queue, number of entries that did not have to wait in the queue, percentage of vehicles not having to wait, average time spent in queue, and the average time in the queue excluding those that didn't have to wait before servicing.

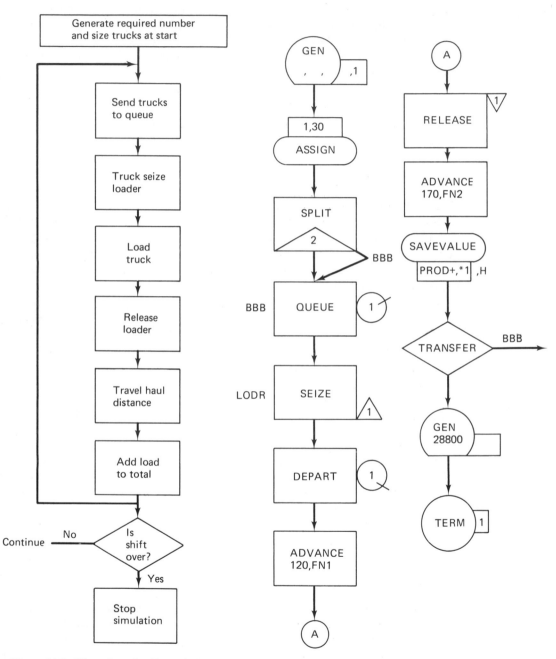

Figure 14-1 Flow chart for Example 14-1.

Figure 14-2 Block diagram for Example 14-1.

Example 14-1

PROBLEM: Set up a GPSS model for a shovel/truck excavating and hauling operation. Use an 8-hour work shift and the data given below. Determine the system production, average shovel utilization, average number of trucks waiting in the queue line, and the average waiting time in the queue for those trucks required to wait.

Truck capacity = 30 LCY.

Mean loading time = 120 sec.

For loading time distribution, use a function modifier described by the following 11 points: (0, 1.75), (0.04, 1.62), (0.07, 1.5), (0.10, 1.4), (0.18, 1.25), (0.28, 1.13), (0.42, 1.01), (0.6, 0.8), (0.89, 0.7), (0.96, 0.55), and (1.0, 0.375).

Mean haul, dump, and return time = 120 sec.

For travel time distribution, use a function modifier described by the following 11 points: (0, 0.39), (0.01, 0.51), (0.04, 0.64), (0.12, 0.76), (0.27, 0.88), (0.5, 1.0), (0.73, 1.12), (0.88, 1.24), (0.96, 1.36), (0.99, 1.48), and (1.0, 1.6).

Solution:

Figure 14-1 illustrates the model flow chart showing only basic program logic. Figure 14-2 shows the GPSS block diagram for the model using GPSS block symbols. Figure 14-3 shows the GPSS program listing and Figure 14-4 illustrates the essential computer output.

Comments on the Solution:

```
BLOCK                                                                    CARD
NUMBER   *LOC   OPERATION A,B,C,D,E,F,G              COMMENTS          NUMBER
                SIMULATE                                                  1
              1 FUNCTION  RN1,C11    ,   ,     ,    ,      ,              2
         0,1.75/.04,1.62/.07,1.5/.10,1.4/.18,1.25/.28,1.13/              3
         .42,1.01/.6,.8/.89,.7/.96,.55/1.0,.375                          4
              2 FUNCTION  RN2,C11 . . . . .                              5
         0,.39/.01,.51/.04,.64/.12,.76/.27,.88/.5,1.0/                   6
         .73,1.12/.88,1.24/.96,1.36/.99,1.48/1.0,1.6                     7
   1            GENERATE  ,,,1                                            8
   2     FFF    ASSIGN    1,30                                            9
   3            SPLIT     2,BBB                                          10
   4     BBB    QUEUE     1                                             11
   5     LODR   SEIZE     1                                             12
   6            DEPART    1                                             13
   7     GGG    ADVANCE   120,FN1                                       14
   8            RELEASE   1                                             15
   9     TTT    ADVANCE   170,FN2                                       16
  10            SAVEVALUE PROD+,*1,H                                    17
  11            TRANSFER  ,BBB                                          18
  12            GENERATE  28800                                         19
  13            TERMINATE 1                                             20
              2 QTABLE    1,0,50,20                                     21
                START     1                                             22
                END                                                     23
```

Figure 14-3 Program listing for Example 14-1.

RELATIVE CLOCK 28800 ABSOLUTE CLOCK 28800
TERMINATIONS TO GO 1

QUEUE	MAXIMUM CONTENTS	AVERAGE CONTENTS	TOTAL ENTRIES	ZERO ENTRIES	PERCENT ZEROS	AVERAGE TIME/TRANS	$AVERAGE TIME/TRANS	TABLE NUMBER	CURRENT CONTENTS
1	2	.574	246	28	11.3	67.272	75.912	2	

$AVERAGE TIME/TRANS = AVERAGE TIME/TRANS EXCLUDING ZERO ENTRIES

TABLE 2

ENTRIES IN TABLE	MEAN ARGUMENT	STANDARD DEVIATION	SUM OF ARGUMENTS	NON-WEIGHTED
246	67.272	52.062	16549.000	

UPPER LIMIT	OBSERVED FREQUENCY	PER CENT OF TOTAL	CUMULATIVE PERCENTAGE	CUMULATIVE REMAINDER	MULTIPLE OF MEAN	DEVIATION FROM MEAN
0	28	11.32	11.3	88.6	-.000	-1.292
50	81	32.92	44.3	55.6	.743	-.331
100	76	30.89	75.2	24.7	1.486	.628
150	45	18.29	93.4	6.5	2.229	1.589
200	13	5.28	98.7	1.2	2.972	2.549
250	2	.81	99.5	.4	3.716	3.509
300	1	.40	100.0	.0	4.459	4.470

REMAINING FREQUENCIES ARE ALL ZERO

FACILITY	AVERAGE UTILIZATION	NUMBER ENTRIES	AVERAGE TIME/TRAN	SEIZING TRANS. NO.	PREEMPTING TRANS. NO.
1	.980	246	114.760	4	

CONTENTS OF HALFWORD SAVEVALUES (NON-ZERO)

SAVEVALUE NR, VALUE NR, VALUE NR, VALUE NR, VALUE NR, VALUE NR, VALUE
PROD 7290

Figure 14-4 Output for Example 14-1.

The meaning and purpose of each card used in the program is described below.

Card No.	Description and Use
1	SIMULATE. This control card is required to initiate a simulation run. If it is omitted, the program will be compiled and checked for errors but no simulation will be run.
2–4	1 FUNCTION. These three cards describe the shape of the modifier curve that is applied to the mean loading time to obtain actual loading times for each truck load. Eleven coordinates (X, Y) are used to describe the modifier curve.
5–7	2 FUNCTION. These three cards describe the modifier distribution curve used for computation of the actual travel time for each truck. Again, 11 coordinates describe this function.
8	GENERATE. This card is used to create one transaction (in this case a truck) which will be the parent truck used for later additions. This one truck is sent into the system.
9	ASSIGN. This is the principal means for describing the vehicles in the system. In this case, the card is used to specify the hauling capacity of each truck (30 LCY).
10	SPLIT. A way to increase the number of trucks in the system (in addition to the GENERATE card) is by the use of a SPLIT card. In this program the split card creates two additional trucks exactly like the parent transaction that entered the block, creating a system containing three trucks.
11	QUEUE. This card signifies that the user expects a queue to exist at this point and desires queue statistics to be recorded. This queue represents trucks waiting to be loaded.
12	SEIZE. This block records the use of the shovel by the entering truck. When a truck is being loaded, other trucks must wait in the queue until the shovel is free.
13	DEPART. When the shovel is free, this card allows one truck to depart from the queue and be serviced by the shovel.
14	ADVANCE. This card specifies the mean time to perform the loading operation. In this case, the mean loading time of 120 sec is adjusted by the modifier described by Function 1.
15	RELEASE. This card is used to release the shovel when loading is completed.
16	ADVANCE. This card specifies the mean haul, dump, and return time. In this example the mean time of 170 sec is modified by the function described by Function 2.

17   SAVEVALUE. This card stores a specified numerical value. In this case, it is the total production of the haul units. Since each truck hauls 30 LCY, the quantity 30 is added to the total when each haul cycle is completed.

18   TRANSFER. This card is used to direct the trucks back to the loading queue at the end of the haul cycle. Trucks continue to be directed to the queue until time runs out.

19   GENERATE. This card is used here as a clock. The 28,800 represents the number of seconds in an 8-hour work day. When this amount of time has been simulated by the program, the simulation stops and all statistics are tabulated.

20   TERMINATE. This card is used to remove transactions from the system and represents the completion of a flow path in the system.

21   2 QTABLE. This card describes the frequency classes for the queue statistics. The width and limits of each class are specified with this card.

22   START. This is a control card that is used to tell the computer to start a simulation.

23   END. This card identifies the end of the program.

Queue statistics (QUEUE 1) indicate that the total number of trucks passing through the queue line was 246. Of these, 28 trucks did not wait at all but moved directly under the shovel. The maximum number of trucks in the queue at any time was 2, as you would expect, since only 3 trucks were in the system and 1 truck would always be under the shovel if all 3 trucks were in the loading pit at the same time. The average time spent in the queue by all trucks (including those trucks with zero time in the queue) was 67.272 sec, while the average waiting time for trucks which did have to stop in the queue was 75.912 sec. Detailed queue statistics are provided in Table 2 in Fig. 14-4.

Statistics for the shovel (FACILITY 1) give average shovel utilization (0.980), total number of trucks loaded during the period (246), and the average loading time (114.760 sec).

The SAVEVALUE output indicates the accumulated value specified by the programmer. In this case, block 2 (card 9) and block 10 (card 17) caused the program to add 30 LCY to the accumulator for each truck loaded and to identify this total on the output as PROD (production). Thus, total system production for the 8-hour day was 7,290 LCY.

Example 14-2

PROBLEM: Revise the program of Example 14-1 to provide 2 shovels for loading, one with a mean loading time of 120 sec and the other with a mean loading time of 140 sec. Use a fleet of seventeen 30 LCY trucks. Simulate a single 8-hour shift.

Solution:

The revised computer program is given in Figure 14-5. The pertinent output is shown in Figure 14-6.

Comments on the Solution:

BLOCK NUMBER	*LOC	OPERATION	A,B,C,D,E,F,G	COMMENTS	CARD NUMBER
		SIMULATE			1
	1	FUNCTION	RN1,C11		2
		0,1.75/.04,1.62/.07,1.5/.10,1.4/.18,1.25/.28,1.13/			3
		.42,1.01/.6,.8/.89,.7/.96,.55/1.0,.375			4
	2	FUNCTION	RN2,C11		5
		0,.39/.01,.51/.04,.64/.12,.76/.27,.88/.5,1.0/			6
		.73,1.12/.88,1.24/.96,1.36/.99,1.48/1.0,1.6			7
1		GENERATE	,,,1		8
2	FFF	ASSIGN	1,30		9
3		SPLIT	16,BBB		10
4	BBB	QUEUE	1		11
5		TRANSFER	BOTH,LDR1,LDR2		12
6	LDR1	SEIZE	1		13
7		DEPART	1		14
8	GGG	ADVANCE	120,FN1		15
9		RELEASE	1		16
10		TRANSFER	,TTT		17
11	LDR2	SEIZE	2		18
12		DEPART	1		19
13		ADVANCE	140,FN1		20
14		RELEASE	2		21
15		TRANSFER	,TTT		22
16	TTT	ADVANCE	170,FN2		23
17		SAVEVALUE	PROD+,*1,H		24
18		TRANSFER	,BBB		25
19		GENERATE	28800		26
20		TERMINATE	1		27
	2	QTABLE	1,0,50,20		28
		START	1		29
		END			30

Figure 14-5 Program listing for Example 14-2.

The program of Example 14-1 is changed as follows. The SPLIT card (card 10) is revised to add 16 trucks to the one initially generated, making a fleet of 17 trucks. The TRANSFER card (card 12) is used to direct arriving trucks to whichever shovel (LDR 1 or LDR 2) is available. Priority is given to the faster shovel. Two sets of SIEZE, DEPART, and ADVANCE cards (cards 13–15 and cards 17–19) are used to control loading operations. The ADVANCE cards (cards 15 and 20) specify the mean loading time and function modifier for each shovel. Since only one queue line is maintained in this situation, only one set of queue statistics is produced. However, the average utilization, average loading time,

RELATIVE CLOCK 28800   ABSOLUTE CLOCK 28800
TERMINATIONS TO GO 1

QUEUE	MAXIMUM CONTENTS	AVERAGE CONTENTS	TOTAL ENTRIES	ZERO ENTRIES	PERCENT ZEROS	AVERAGE TIME/TRANS	$AVERAGE TIME/TRANS	TABLE NUMBER	CURRENT CONTENTS
1	15	12.316	471	2	.4	753.125	756.336	2	11

$AVERAGE TIME/TRANS = AVERAGE TIME/TRANS EXCLUDING ZERO ENTRIES

TABLE 2

ENTRIES IN TABLE 460	MEAN ARGUMENT 762.560	STANDARD DEVIATION 108.562	SUM OF ARGUMENTS 350778.000	NON-WEIGHTED

UPPER LIMIT	OBSERVED FREQUENCY	PER CENT OF TOTAL	CUMULATIVE PERCENTAGE	CUMULATIVE REMAINDER	MULTIPLE OF MEAN	DEVIATION FROM MEAN
0	2	.43	.4	99.5	-.000	-7.024
50	0	.00	.4	99.5	.065	-6.563
100	1	.21	.6	99.3	.131	-6.103
150	0	.00	.6	99.3	.196	-5.642
200	0	.00	.6	99.3	.262	-5.181
250	1	.21	.8	99.1	.327	-4.721
300	2	.43	1.3	98.6	.393	-4.260
350	0	.00	1.3	98.6	.458	-3.800
400	1	.21	1.5	98.4	.524	-3.339
450	0	.00	1.5	98.4	.590	-2.879
500	1	.21	1.7	98.2	.655	-2.418
550	1	.21	1.9	98.0	.721	-1.957
600	4	.86	2.8	97.1	.786	-1.497
650	25	5.43	8.2	91.7	.852	-1.036
700	49	10.65	18.9	81.9	.917	-.576
750	111	24.13	43.0	56.9	.983	-.115
800	91	19.78	62.8	37.1	1.049	.344
850	94	20.43	83.2	16.7	1.114	.805
900	56	12.17	95.4	4.5	1.180	1.265

OVERFLOW 21   4.56   100.0   .0

AVERAGE VALUE OF OVERFLOW 918.38

FACILITY	AVERAGE UTILIZATION	NUMBER ENTRIES	AVERAGE TIME/TRAN	SEIZING TRANS. NO.	PREEMPTING TRANS. NO.
1	.999	252	114.281	6	
2	.999	208	138.456	14	

CONTENTS OF HALFWORD SAVEVALUES (NON-ZERO)
SAVEVALUE   NR, VALUE   NR, VALUE   NR, VALUE   NR, VALUE   NR, VALUE   NR, VALUE   NR, VALUE   NR, VALUE
PROD 13620

Figure 14-6 Output for Example 14-2.

and the number of trucks loaded are shown separately for each shovel (FACILITY 1 and FACILITY 2). A single value (labeled PROD) indicating total fleet production in loose cubic yards (13,620) is tabulated.

Example 14-3

PROBLEM: Write a GPSS program to simulate a push-pull scraper operation. Use the mean loading time (which in this case represents hook up time plus the time required to load both scrapers) and function modifier of Example 14-1. Use the mean hauling time and its function modifier from Examples 14-1 and 14-2.

Solution:

The GPSS program listing is given in Figure 14-7 and the program output is shown in Figure 14-8.
Comments on the Solution:

```
BLOCK                                                          CARD
NUMBER   *LOC     OPERATION A,B,C,D,E,F,G          COMMENTS    NUMBER
                  SIMULATE                                     1
               1 FUNCTION   RN1,C11      .   .      .   .  .   2
         0,1.75/.04,1.62/.07,1.5/.10,1.4/.18,1.25/.28,1.13/   3
         .42,1.01/.6,.8/.89,.7/.96,.55/1.0,.375               4
               2 FUNCTION   RN2,C11      .   .      .   .      5
         0,.39/.01,.51/.04,.64/.12,.76/.27,88/.5,1.0/         6
         .73,1.12/.88,1.24/.96,1.36/.99,1.48/1.0,1.6          7
1                 GENERATE   ,,,1                              8
2        FFF      ASSIGN     1,30                              9
3                 SPLIT      7,BBB                             10
4        BBB      QUEUE      1                                 11
5                 ASSEMBLE   2                                 12
6        LODR     SEIZE      1                                 13
7                 DEPART     1,2                               14
8        GGG      ADVANCE    120,FN1                           15
9                 RELEASE    1                                 16
10                SPLIT      1,TTT                             17
11       TTT      ADVANCE    170,FN2                           18
12                SAVEVALUE  PROD+,*1,H                        19
13                TRANSFER   ,BBB                              20
14                GENERATE   14400                             21
15                TERMINATE  1                                 22
               2 QTABLE      1,0,50,20                         23
                  START      1                                 24
                  END                                         25
```

Figure 14-7 Program listing for Example 14-3.

This program differs from Examples 14-1 and 14-2 in that two scrapers must be present at the loading area before loading can begin. Thus, the first scraper to arrive in the cut area must wait until a second scraper arrives and hooks up for loading. This action is simulated by the use of the ASSEMBLE card described below. After loading is completed, the scrapers uncouple and proceed individually to perform the haul, dump, and return portions of the cycle. The uncoupling

RELATIVE CLOCK 14400   ABSOLUTE CLOCK 14400
TERMINATIONS TO GO 1

QUEUE	MAXIMUM CONTENTS	AVERAGE CONTENTS	TOTAL ENTRIES	ZERO ENTRIES	PERCENT ZEROS	AVERAGE TIME/TRANS	$AVERAGE TIME/TRANS	TABLE NUMBER	CURRENT CONTENTS
1	8	3.115	250	2	.7	179.427	180.875	2	2

$AVERAGE TIME/TRANS = AVERAGE TIME/TRANS EXCLUDING ZERO ENTRIES

TABLE 2

ENTRIES IN TABLE	MEAN ARGUMENT	STANDARD DEVIATION	SUM OF ARGUMENTS	
124	198.258	69.875	24584.000	NON-WEIGHTED

UPPER LIMIT	OBSERVED FREQUENCY	PER CENT OF TOTAL	CUMULATIVE PERCENTAGE	CUMULATIVE REMAINDER	MULTIPLE OF MEAN	DEVIATION FROM MEAN
0	1	.80	.8	99.1	-.000	-2.837
50	0	.00	.8	99.1	.252	-2.121
100	8	6.45	7.2	92.7	.504	-1.406
150	26	20.96	28.2	71.7	.756	-.690
200	26	20.96	49.1	50.8	1.008	.024
250	30	24.19	73.3	26.6	1.260	.740
300	26	20.96	94.3	5.6	1.513	1.456
350	6	4.83	99.1	.8	1.765	2.171
400	0	.00	99.1	.8	2.017	2.887
450	1	.80	100.0	.0	2.269	3.602

REMAINING FREQUENCIES ARE ALL ZERO

FACILITY	AVERAGE UTILIZATION	NUMBER ENTRIES	AVERAGE TIME/TRAN	SEIZING TRANS. NO.	PREEMPTING TRANS. NO.
1	.999	124	116.120	7	

CONTENTS OF HALFWORD SAVEVALUES (NON-ZERO)
SAVEVALUE   NR, VALUE   NR, VALUE   NR, VALUE   NR, VALUE   NR, VALUE   NR, VALUE   NR, VALUE
PROD   7260

Figure 14-8 Output for Example 14-3.

operation is simulated by use of the SPLIT card described below. Note that the same distribution of loading and hauling times that were used in Examples 14-1 and 14-2 are assumed here. Since these time distributions were based on observations of truck/shovel operations, it is doubtful that they are completely valid for this case. Time studies of push-pull scraper operations should be performed to develop better time distribution curves for the push-pull model.

Important GPSS cards used in this program and their functions are as follows:

Card No.	Comment
9	The capacity of *each* scraper.
10	Creates seven additional scrapers, making a total of eight. Note that the total number of scrapers created by the GENERATE and its succeeding SPLIT card should be a multiple of two since scrapers operate in pairs.
12	The ASSEMBLE card is used to mate two scrapers for loading.
14	Two units leave the queue when they hook up for loading.
15	The mean loading time includes hookup time plus loading time for *both* scrapers.
17	The SPLIT card separates the two scrapers upon completion of loading so that they can operate individually for hauling, dumping, and returning.
18	The mean scraper haul time and function modifier is specified in this ADVANCE card. This time distribution applies to every individual scraper leaving the loading area.

Queue statistics, facility utilization (0.999), and total production (7,260 LCY) are provided in the program output.

PROBLEMS

1. Explain the concept of equipment selection from a systems viewpoint.
2. Draw a simple flow diagram representing a simulation program for a push-pull scraper operation.
3. Draw a simple flow diagram for the program of Example 14-2.
4. Draw a GPSS block diagram for the program of Example 14-2.
5. Write the GPSS FUNCTION and ADVANCE instructions which describe a loading operation having a triangular time distribution curve with a mean value of 100 sec, a minimum value of 40 sec, and a maximum value of 140 sec.

REFERENCES

1. *Capital Investment Studies Using GPSS: Bulk Material Movement Problems.* White Plains, New York: International Business Machines Corporation, 1968.
2. *Earthmoving System Selection.* Peoria, Illinois: Caterpillar Tractor Co., n.d.

3. *General Purpose Simulation System/360: Introductory Users Manual*. White Plains, New York: International Business Machines Corporation, 1969.

4. GORDON, GEOFFREY, *Application of GPSS V to Discrete System Simulation*. Englewood Cliffs, New Jersey: Prentice-Hall, 1975.

5. GREENBERG, STANLEY, *GPSS Primer*. New York: Wiley-Interscience, 1972.

6. LEWIS, D. A., and W. C. MORGAN, "Tractor-Scraper Performance Evaluation Using Digital Computer," Society of Automotive Engineers, New York: 1961.

7. LEWIS, W. A., and A. W. PARSONS, *The Application of Belt Conveyors in Road Earthworks*. London: The Institution of Civil Engineers, 1973.

8. O'SHEA, J. B., *An Application of the Theory of Queues to the Forecasting of Shovel-Truck Fleet Production*, Civil Engineering Studies, Series No. 3. Urbana, Illinois: University of Illinois, 1964.

9. PARSONS, A. W., and B. A. BROAD, *Belt Conveyors in Road Construction: RRL Report LR 33T*. Crowthorne, Berkshire, England: Road Research Laboratory, 1970.

15

PLANT LAYOUT

15-1 PRINCIPLES OF PLANT LAYOUT

Introduction

The design and layout of plants performing repetitive construction operations—whether they be concrete prefabricating yards, plants for fabricating structural assemblies, or concrete, asphalt, or rock crushing plants—constitute important phases of construction planning. In addition to their application to plant layout planning, the methods analysis techniques discussed in Section 15-2 can be used advantageously to improve the performance of any repetitive construction task, whether in a plant or not.

In planning a construction plant operation there are three principal areas of consideration: equipment selection, site selection, and plant layout. These areas are closely interrelated since a decision on equipment selection may strongly influence plant layout and vice versa. Factors to be considered in each of these areas are discussed in the following paragraphs.

Equipment Selection

Principal factors in selecting equipment for a construction plant include the following:

1. Work To Be Accomplished. What specific processes must be performed? How many units or tons must be processed per unit of time? The required

261

capacity of the plant may indicate consideration of one large piece of processing equipment versus several smaller pieces of equipment operating in parallel.

2. Equipment Available. The objective again is to maximize profit on the operation. However, time or capital limitations may limit the choice of available equipment. Even though capital is available, if the most suitable piece of equipment is not a standard item or is in short supply, the time required to obtain the equipment may be excessive.

3. Raw Materials Involved. Consider the type, size, and quantity of raw materials to be processed. Do they require special handling equipment, protective equipment, or other safety measures? Where are the raw materials located and how will they be transported to the plant?

4. Material Flow. Plan the flow of material through the plant. Use the methods described in Section 15-2. Determine the amount, size, and type of equipment needed to load and unload haul units, maintain stockpiles and storage yards, etc.

Site Selection

The choice of the plant site may have already been made or may be dictated by special conditions, such as zoning regulations, etc. When possible, however, the following factors should be considered:

1. Location of Input and Output. When a number of acceptable sites exist, the site should be located so as to minimize transportation costs. In locating a concrete mix plant for a paving operation, for example, the primary criteria may be to minimize the total transportation requirement for raw materials and plastic concrete.

2. Terrain. The nature of the site terrain will often affect the cost of constructing the plant and its supporting facilities.

3. Access and Haul Roads. The plant must be accessible to transportation facilities to be used to bring in raw materials and haul away the finished product. The quality and capacity of highways, railroads, waterways, or airports may limit the size of transport vehicles used and directly influence transportation cost and capacity.

4. Utilities. The availability and cost of utilities are important considerations. When commercial electricity is not available, diesel-powered equipment may be required or electricity may have to be generated on site. Water supply is often critical, particularly when required in large quantities for washing raw materials or for cooling purposes.

Plant Layout

In planning the plant layout itself the techniques of Section 15-2 should be used to minimize the movement of material, equipment, and personnel as well as to minimize processing time. However, other considerations that may be of importance are discussed below.

1. Drainage. Typically, construction plants are set up on or near construction project sites. The site topography and nature of the soil, as well as climatic conditions, will determine whether or not drainage will be a problem. The lack of adequate drainage may greatly hinder plant operations. If extensive work is required to provide adequate drainage, an analysis should be made of the cost of such work versus losses to be expected because of lack of drainage.

2. Prevailing Winds. Plants that exhaust particulate matter or gases or that contain equipment sensitive to dust should consider the prevailing winds and nature of the surroundings. Rock-crushing plants, for example, should be located so that the wind carries dust away from the processing equipment. Supporting facilities such as generators, offices, shops, etc., should be located out of the path of the dust flow.

3. Construction Aids. The use of construction aids, such as jigs, in repetitive construction processes will often be beneficial. However, the cost and time to fabricate construction aids must be weighed against the expected reduction in plant processing time and cost.

4. Material Handling and Storage. Adequate space must be provided for handling and storing both raw materials and finished products. Covered storage may be required to protect material from the weather.

5. Other Facilities. Don't overlook the requirements for supporting facilities, such as offices, equipment maintenance, parking space for workers, workers' facilities for rest, eating, and sanitation, emergency medical facilities, water storage, lighting for security and night operations, etc.

6. Road Network. Whenever possible, separate road networks should be provided for incoming materials, outgoing products, and internal service access. Any roads constructed must be designed to adequately support anticipated loads. If paved roads are not to be provided, provision must be made for adequate road maintenance. Traffic flow and speed limits must be coordinated with the plant layout while minimizing safety hazards.

15-2 METHODS ANALYSIS

Introduction

Methods analysis (also known as *methods improvement, methods engineering, work improvement,* or *work simplification*) is the process of analyzing a task and devising a method of accomplishing the task which will minimize the time and effort required to perform the task. In spite of efforts to introduce such methods into construction early in the twentieth century, these methods have still not been widely adopted in the construction industry. The manufacturing industry, however, has readily adopted such methods and has reaped the benefits of increased productivity and reduced manufacturing costs. When such techniques have been applied to construction projects, however, they have often been highly successful.

In applying methods analysis to the design and layout of construction plants, we will discuss the use of four charts or diagrams: *flow diagrams* (or flow charts), *flow process charts*, *gang process charts*, and *layout sketches*. The example discussed and Figures 15-1 through 15-6 come from Reference 5 and U.S. Army Engineer School instructional aids. Standard symbols which have been adopted for use in flow and process charts are illustrated in Figure 15-1.

"Do" operations, or main steps in a process	○	Operation	Cutting wood, digging a ditch, or positioning a part
An operation which changes the shape of the material	●	Operation	Making a cut in a board
Movement of material from one place to another	⇨	Transportation	Carrying a truss or a unit
Verifies quality, quantity, or approval	☐	Inspection	Checking a center line or testing equipment
Delay, awaiting completion of an interrelated job	D	Delay	Awaiting use
Keep	▽	Storage	Material in storage

Figure 15-1   Flow chart symbols. (U.S. Department of the Army.)

Flow Diagram

The flow diagram for a plant serves the same purpose as a flow diagram for a computer program. That is, it traces the flow of materials through a series of processing steps. Its use should enable the planner to visualize the entire operation and determine the total number of operations, material movements, and delays involved in the operation. The flow diagram should then be analyzed and revised to reduce the number of operations, movements, and delays to a minimum. The flow diagram does not indicate how far materials are moved, the time required for each step, or the number of workers required. Hence, the flow diagram should be used to make a preliminary plan of operation which may be revised as flow process charts and gang process charts are made and analyzed. As the flow diagram is developed, a preliminary layout of storage areas, equipment, and operating areas will be conceived. A rough sketch of this layout should be made for later use in developing the detailed layout sketch.

To prepare a flow diagram, first list all of the major process or work stations down the left side of the diagram. Next, use the symbols of Figure 15-1 to indicate the major tasks performed at each station. Finally, connect the symbols to illustrate the flow of work through the plants.

An operation for prefabricating the timber roof truss shown in Figure 15-2 will be analyzed by the use of the methods named above. Figure 15-3 illustrates the second flow diagram prepared for the fabrication operation. The first flow diagram prepared (not shown) required a total of 95 operations, movements storages, and delays using one saw on the precutting line for the truss members. Since all members of the truss except the lower chord splice require two saw cuts at different angles, the total number of steps is reduced considerably (95 to 80) by using two saws in the precutting operation. Note the use of a construction aid (jig) in the assembly process.

20 ft. span roof trusses

Item no.	Member	Pcs per truss	Length in place	Size	Comm. length	Pcs per length	Saw setups
1	Rafter	2	10′ – 11″	2″ x 4″	12′	1	2
2	Lower chord	2	10′ – 0″	1″ x 6″	10′	1	2
3	Web	2	5′ – 5″	2″ x 4″	12′	2	2
4	Hangar	1	$4′ - 4\frac{1}{2}″$	1″ x 6″	14′	3	2
5	Rafter tie	1	1′ – 8″	1″ x 8″	12′	8	2
6	Lower chord splice	1	2′ – 6″	2″ x 6″	10′	4	1

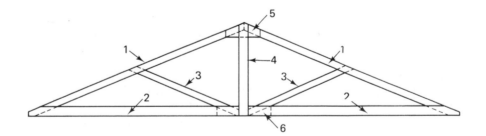

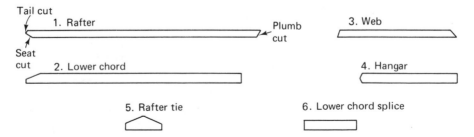

Tail cut
1. Rafter
Plumb cut
Seat cut
2. Lower chord

3. Web
4. Hangar

5. Rafter tie
6. Lower chord splice

Figure 15-2 Roof truss components. (U.S. Department of the Army.)

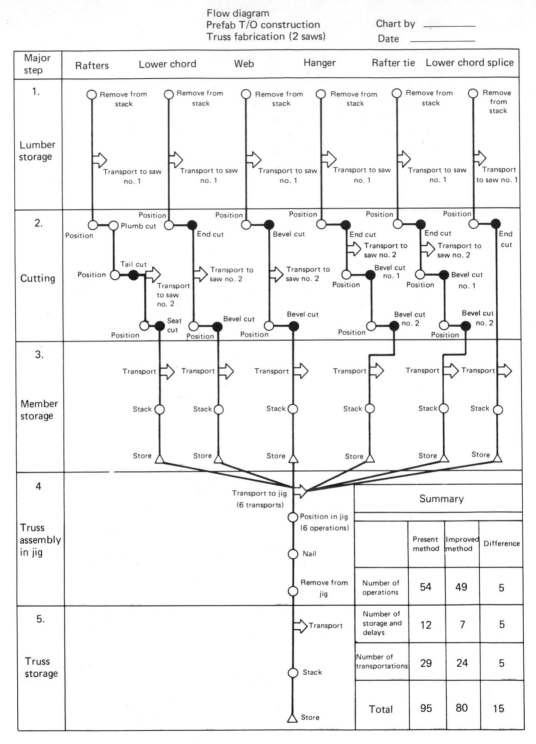

Flow diagram
Prefab T/O construction
Truss fabrication (2 saws)

Chart by _____
Date _____

Major step											
	Rafters	Lower chord	Web	Hanger	Rafter tie	Lower chord splice					

Summary

	Present method	Improved method	Difference
Number of operations	54	49	5
Number of storage and delays	12	7	5
Number of transportations	29	24	5
Total	95	80	15

Figure 15-3 Flow diagram for constructing roof trusses. (U.S. Department of the Army.)

266

Flow Process Chart

After the flow diagram has been completed, a flow process chart is prepared to analyze each major component of the flow diagram. In this chart the distance and time required for each transportation and the time required for each operation and delay are estimated.

List in the left-hand column a brief description of each activity as it occurs. Trace the work flow by connecting the appropriate symbols in the second column. Enter the transportation distance, number of times the activity is repeated, and its estimated time. The time required to perform the entire operation shown on the chart is determined by the total time needed to perform the steps that cannot be performed concurrently. This time is called the *control factor*. The control factor can only be reduced by speeding up its activities or by devising a method that allows some of its activities to be performed concurrently.

Figure 15-4 shows a flow process chart for the operation of precutting rafters for the roof truss using two saws in the process. The comment in the Notes column indicates that the control factor is now 8 sec instead of the 12 sec previously required when using one saw. Remember that two rafters are required per truss. Thus, the expected rafter production per 50-minute hour, using an efficiency factor of 70%, is calculated at 131 sets of rafters per hour.

As the flow process charts are prepared for each element of truss fabrication, additional details of plant layout will be developed. These should be incorporated in the layout sketch previously started.

Layout Sketch

After the flow process charts have been analyzed, additional effort may be placed on improving the plant layout. Some of the principles of plant layout have been discussed previously. Remember that the major objective in plant layout is to minimize in-plant processing time and effort. Thus, the number of material movements and the distance between operations must be minimized. The delivery of raw materials should be carefully scheduled in order to reduce storage requirements consistent with maintaining an operating level that will compensate for any expected delivery delays. Rehandling of material and in-plant storage should be minimized. Always keep safety in mind. Avoid traffic flow patterns that will present safety hazards or result in traffic delays.

Figure 15-5 illustrates the final layout sketch for the truss prefabricating operation. Notice that the position of workers is indicated on the sketch. These positions were developed from the flow process charts and the gang process chart described below. The final layout sketch is the result of a careful analysis of the operation to be performed and its use will enable a supervisor to establish and operate the plant with a minimum of additional guidance.

| FLOW PROCESS CHART (DA Pamphlet 20-300) #1 | | NUMBER J.O.0000 | | PAGE NO. 1 | | NO. OF PAGES 1 |

FLOW PROCESS CHART
(DA Pamphlet 20-300) #1

NUMBER	PAGE NO.	NO. OF PAGES
J.O.0000	1	1

PROCESS
Cutting Rafters
☐ MAN OR ☒ MATERIAL

SUMMARY						
ACTIONS		PRESENT		PROPOSED		DIFFERENCE
		NO.	TIME	NO.	TIME	NO.	TIME
○ OPERATIONS				8	18	
⇨ TRANSPORTATIONS				3	16	
☐ INSPECTIONS				0	0	
D DELAYS				0	0	
▽ STORAGES				1		

CHART BEGINS	CHART ENDS
At Lumber Storage	At Parts Storage

CHARTED BY	DATE
Capt Doit	2 July 65

ORGANIZATION
477th Engr Const Bn

DISTANCE TRAVELLED *(Feet)* 53 ft

#	DETAILS OF ☐ PRESENT / PROPOSED METHOD	OPERATION / TRANSPORTATION / INSPECTION / DELAY / STORAGE	DISTANCE IN FEET	QUANTITY	TIME	ANALYSIS WHY? (WHAT? WHERE? WHEN? WHO? HOW?)	NOTES — Time in seconds.	ANALYSIS CHNGE (ELIMINATE / COMBINE / SEQUENCE / PLACE / PERSON / IMPROVE)
1	Remove lumber from storage	○⇨☐D▽		1	3			
2	Transport to saw table	○⇨☐D▽	15	1	6			
3	Position for plumb cut	○⇨☐D▽		1	2			
4	Make plumb cut	●⇨☐D▽		1	2			
5	Position for tail cut	○⇨☐D▽		1	2			
6	Make tail cut	●⇨☐D▽		1	2			
7	Trans to 2d saw	○⇨☐D▽	14	1	4		With 2 saws	
8	Position for seat cut	○⇨☐D▽		1	2		Control factor is	
9	Make seat cut	●⇨☐D▽		1	2		reduced from	
10	On conveyor belt to storage	○⇨☐D▽	24	1	6		12 sec to 8 sec	
11	Remove fr convey stack	○⇨☐D▽		1	3		because operations	
12	Rafter in storage.	○⇨☐D▽					run concurrently.	
13		○⇨☐D▽						
14		○⇨☐D▽					Production Rate =	
15		○⇨☐D▽					one (1) rafter each	
16		○⇨☐D▽					eight (8) sec or	
17		○⇨☐D▽					16 sec/truss unit.	
18		○⇨☐D▽					Assume 70% eff	
19		○⇨☐D▽					work hour 50 min	
20		○⇨☐D▽					$\dfrac{60 \times 50 \times .70}{16} = \dfrac{2100}{16}$	
21		○⇨☐D▽					131 Truss units/hr	

DA FORM 1 MAY 51 **684** *(Formerly DA AGO)* REPLACES OCS FORM 391, 1 FEB 1951, WHICH MAY BE USED. 3

Figure 15-4 Flow process chart for cutting rafters. (U.S. Army Engineer School.)

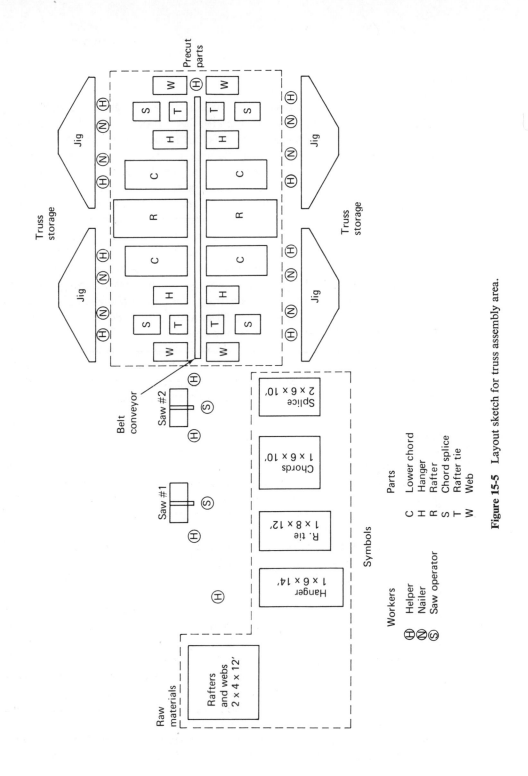

Figure 15-5 Layout sketch for truss assembly area.

Gang Process Chart

A gang process chart documents the activities of a group of workers (gang or crew) in a manner similar to that used for flow process charts. It also utilizes the symbols of Figure 15-1. The gang process chart is used to analyze the crew organization and work performed by each member of the crew. It should represent a complete cycle for the crew member who accomplishes the greatest number of steps. Steps that do not occur often enough to be significant may be omitted.

Figure 15-6 illustrates a gang process chart for the truss assembly phase of the roof truss plant operation. Note that the sequence numbers and descriptions in the right-hand column do not correspond with the step occupying the same horizontal row but are keyed to the numbers placed in the symbols. Thus, the third row or step indicates that the two helpers return to the precut storage area while the two nailers are idle. In the fourth step one helper carries a chord splice to the assembly table while the other helper carries a hangar to the assembly table. Based on the total cycle time of 180 sec, a 50-minute hour, and 70% efficiency, 11.2 trusses per hour can be assembled by one crew.

A crew balance chart is similar to a gang process chart in that it presents information on the activities of each crew member but does so in a graphical form. The crew balance chart uses a vertical bar for each worker which is divided into blocks to indicate the percentage of the worker's time devoted to each activity.

Improving Existing Operations

Existing construction operations may also be analyzed by the methods described above. In order to accomplish this, the operation of each worker or machine must be observed and timed. The two principal methods for observing and recording the details of existing methods are time studies (or stopwatch studies) and time-lapse photography.

In a time study a record is made of each activity performed and the time required for its performance. A stopwatch is usually used in timing such operations; hence, the name stopwatch study. The term *time-motion study* is sometimes applied to such studies. However, a true time-motion study is a detailed analysis of the movements (usually hand motions) of a worker. The methods analysis studies conducted on construction work do not normally extend to such a level of detail of worker performance.

In recent years equipment has been developed to utilize time-lapse photography for studies of construction work. This equipment usually consists of special models of 8 mm movie cameras capable of exposing one frame at any desired interval. Speeds frequently used range from one exposure per second to one exposure in 4 sec. At a speed of one exposure per 4 sec one Super 8 roll will record 4 hr of operation. Projection equipment used with time-lapse photography allows the viewer to observe the action at any desired speed. Thus, the operation

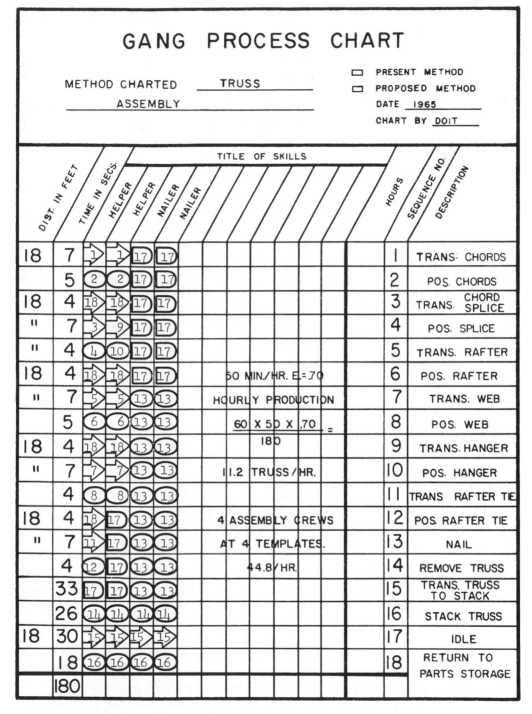

Figure 15-6 Gang process chart for truss assembly. (U.S. Army Engineer School.)

may be viewed as long as necessary to obtain all the desired information. Flow process charts, crew balance charts, etc., of existing operations may be prepared while the time-lapse films are being viewed.

Television cameras and video recorders have also been developed to perform time-lapse studies. Their advantages include the ability to reuse the recording tape and the elimination of film processing time. However, the equipment cost and the cost of recording tape are relatively high.

PROBLEMS

1. You are selecting a location for a continuous flow asphalt plant to provide the hot mix to be used in paving a new 10-mile stretch of interstate highway. You have calculated that approximately 106,000 tons of hot mix will have to be produced. Describe the principal factors you will consider in selecting the plant site.

2. Four sites are under consideration for location of a central mix concrete plant to produce approximately 55,000 cu yd of concrete to be used in paving a new airport runway. Sites under consideration and haul distances for materials involved are as follows:

			Haul Distance (miles)			
Site	Location	To Center of Runway	Coarse Aggregate	Sand	Cement	Water
A	Sand pit	7.0	10.0	0.0	15.0	0.0
B	Rail siding	10.0	15.0	15.0	0.0	2.0
C	Quarry	3.0	0.0	10.0	15.0	0.0
D	Near runway	0.2	3.2	6.8	9.8	1.0

The concrete mix design per cubic yard is as follows:

Coarse aggregate	2,150 lb
Sand	1,100 lb
Water	275 lb
Cement	560 lb

Find the total hauling requirements in ton-miles for each site. What is the optimum plant location based on hauling requirements?

3. Draw a layout diagram of the concrete plant described in Problem 2 above showing the location of stockpiles, the layout of plant components, access roads required, and the traffic pattern.

4. Prepare a flow process chart for the process of cutting the lower chords for the roof truss based on Figures 15-2 and 15-3. Use transportation speeds and operation times based on the values used in Figure 15-4.

5. Analyze the work performed by each member of the crew shown in Figure 15-6 in terms of operation time, transportation time, and idle time. Can you suggest any improvements to the crew organization or work methods?

REFERENCES

1. BARNES, RALPH M., *Motion and Time Study*. New York: John Wiley & Sons, 1968.

2. KRICK, EDWARD V., *Methods Engineering*. New York: John Wiley & Sons, 1962.

3. LEHRER, ROBERT N., *Work Simplification*. Englewood Cliffs, New Jersey: Prentice-Hall, 1957.

4. PARKER, HENRY W., and CLARKSON H. OGLESBY, *Methods Improvement for Construction Managers*. New York: McGraw-Hill, 1972.

5. *TM 5-333: Construction Management*. Washington, D.C.: U.S. Department of the Army, 1972.

INDEX